WITHDRAWN FROM STOCK

Active filters for communications and instrumentation

Active filters for communications and instrumentation

P. Bowron
University of Bradford, England

and

F. W. Stephenson
Virginia Polytechnic Institute and State University, USA

McGRAW-HILL Book Company (UK) Limited

London · New York · St Louis · San Francisco · Auckland
Bogotá · Guatemala · Hamburg · Johannesburg · Lisbon
Madrid · Mexico · Montreal · New Delhi · Panama · Paris · San Juan
São Paulo · Singapore · Sydney · Tokyo · Toronto

Published by
McGRAW-HILL Book Company (UK) Limited
MAIDENHEAD · BERKSHIRE · ENGLAND

British Library Cataloguing in Publication Data

Bowron, P.
Active filters for communications and instrumentation.
1. Electric filters, Active
I. Title II. Stephenson, F.W.
621.3815′32   TK7872.F5   78-40978

ISBN 0-07-084086-5
ISBN 0-07-084085-7   Pbk

Copyright © 1979 McGraw-Hill Book Company (UK) Limited.
All rights reserved. No part of this publication may be reproduced, stored in a retrieval system, or transmitted, in any form or by any means, electronic, mechanical, photocopying, recording, or otherwise, without the prior permission of McGraw-Hill Book Company (UK) Limited.

1 2 3 4 5 SWB 81079

PRINTED AND BOUND IN GREAT BRITAIN

Contents

Acknowledgements — vii

Chapter 1. Introduction — 1
  1.1 The influence of advancing technology — 2
  1.2 Applications of active filters — 2
  1.3 The need for the text — 4
  1.4 Outline of contents — 5
  1.5 Special features — 7

Chapter 2. The Approximation Problem — 9
  2.1 Magnitude approximations — 10
  2.2 Delay and related approximations — 19
  2.3 Frequency transformations — 21

Chapter 3. Passive Networks — 26
  3.1 Properties of immittance functions — 26
  3.2 Transformations and duality — 29
  3.3 Realizability — 32
  3.4 Synthesis procedures — 37
  3.5 Distributed-parameter networks — 47
  3.6 Limitations of passive networks — 50

Chapter 4. Selectivity, Sensitivity, and Stability — 52
  4.1 Selectivity — 52
  4.2 Single-parameter sensitivity — 59
  4.3 Multiparameter sensitivity — 76

Chapter 5. Active Elements — 85
  5.1 Pathological elements — 85
  5.2 Operational amplifiers — 87
  5.3 Controlled sources — 89
  5.4 Immittance converters and inverters — 104
  5.5 Perspective — 122

Chapter 6.	Controlled-Source Realizations		125
	6.1	Single-source methods	125
	6.2	Dual-source methods	138
	6.3	Mixed-source methods	147
	6.4	Optimum design and sensitivity	157
Chapter 7.	Realization using Single Operational Amplifiers		163
	7.1	Single-feedback circuits	163
	7.2	Multiple-loop feedback circuits	168
	7.3	Differential-input amplifier circuits	172
	7.4	Optimum design and sensitivity	179
Chapter 8.	Realization by Analogue Simulation		185
	8.1	Direct replacement of inductors	185
	8.2	Analogue methods	193
	8.3	GIC methods	202
	8.4	Optimum design and sensitivity	212
Chapter 9.	Practical Considerations		220
	9.1	Component fabrication technologies	220
	9.2	Device imperfections	228
	9.3	Environmental susceptibility	239
	9.4	Performance comparisons	241
	9.5	Tuning	247
	9.6	High-order filters	251
	9.7	Commercial availability	254
	9.8	Future developments	257

Appendices

1.	Normalization and denormalization	263
2.	Predistortion	265
3.	Nodal admittance matrix	267
4.	Signal-flow graphs	270
5.	Matrix conversion table	272
6.	Sample design	273

Index   280

Acknowledgements

The authors wish to express their appreciation to the many people who assisted in the preparation of this text as well as to the multitude of past colleagues and co-workers in the field whose inspiration contributed to the climate of thought from which this book has emerged. Though it is impossible to mention everyone by name, our gratitude is nonetheless sincere.

The Universities of Bradford and Hull, as well as Virginia Polytechnic Institute and State University, are to be thanked for the provision of facilities throughout the three years of manuscript preparation. Several colleagues advised in specific areas; in particular, at Hull, Dr J. Dunning-Davies rendered some mathematical assistance on sensitivity optimization in Chapter 6, and Dr J. I. Sewell offered general comments on content as well as guidance on circulators while, at Bradford, occasional network analysis was performed by Messrs M. A. Mohamed, C. Tapinos, A. S. S. Al-Kabbani and C. Lappas. Acknowledgement is also due to the UK Science Research Council for financial support of projects from which some of the results herein are taken. Valuable criticism of the manuscript was provided by a former colleague of the authors, Dr M. R. Lee of Newcastle upon Tyne Polytechnic.

Dr F. W. Stephenson is indebted to the R. T. French Company for funding as Visiting Professor to the University of Rochester, New York, USA during the academic year 1976–77. There, a light teaching load and extensive library facilities enabled him to concentrate on writing sections of the text. Particular thanks are extended to Professor C. W. Merriam, Chairman of the Electrical Engineering Department at Rochester for making available a travelling allowance and to Miss Marybeth Dunn for her excellent secretarial assistance.

Many companies provided data, samples, and photographs. A number of visits to manufacturing plant representative of the various technologies and filter types furnished valuable information for inclusion in later chapters. In particular, we are grateful to the following: Messrs R. Adams, R. J. Dow, C. F. Kurth, and Dr R. A. Friedenson, all of Bell Telephone Laboratories, North Andover, Massachusetts; Dr R. W. Steer, President of Frequency Devices Inc., Haverhill, Massachusetts, for his advice on sections of Chapter 9; Messrs D. Kohn and S. Thanos of Rockland Systems Corporation, West Nyack, New

York; Messrs R. Huntzinger, B. Walmsley, and H. Whitlinger of RCA Applications Division, Somerville, New Jersey. In the UK, valuable advice and assistance has been afforded by Professor P. L. Kirby, Director of Research at Welwyn Electric Ltd, Bedlington, Northumberland.

Finally, we wish to record our sincere appreciation to Mrs Sally Stephenson who, as well as tolerating life with a prospective author, found the time and energy to type most of the manuscript, the assistance of Mrs Sharon Andrews and Mrs Ricki Connor being gratefully acknowledged for the remainder.

P. Bowron
F. W. Stephenson

January 1978

1. Introduction

Modern communications and instrumentation systems are demanding ever more stringent filtering specifications. This calls for a comprehensive treatment of associated technologies, for the purpose of this text, a *filter* being defined[1] as: 'a transducer for separating waves on the basis of their frequencies'.

It is possible to classify electric-wave filters in terms of the reactive elements which they include. These depend[2] on the frequency of operation. For example, resonant cavities are suitable at microwave frequencies, while piezoelectric crystal units provide excellent selectivity in the 5 to 150 MHz band, though with a somewhat limited functional versatility generally confined to bandpass and bandstop characteristics. Though less selective, ceramic filters meet close specifications in the range between 100 kHz and 10 MHz, covering the intermediate frequencies of radio reception. In the audio band down to about 0.1 Hz, mechanical filters are capable of medium to high selectivity with good tunability but offset by a limited function range and difficulties of miniaturization at low frequencies.

The remaining gap in the frequency spectrum between 20 kHz and 100 kHz is, for most purposes, adequately filled by filter networks comprising lumped inductors and capacitors. With some deterioration in performance, the range of application of the latter can be extended to overlap with both crystal and mechanical filters. The principal advantage of passive LCR networks lies in the availability of exact synthesis techniques which facilitate design. However, in practice, the components are lossy and prone to electromagnetic coupling. Most important, from the point of view of incorporation in large-scale systems, is their excessive physical size for low cut-off frequencies.

There is thus a requirement for more compact filtering, particularly in the lower and sub-audio bands. This can be met by including active elements to effectively replace the inductors, thereby introducing the subject of this book: namely, RC-active filters. These allow savings in size and weight even at frequencies approaching 1 MHz and provide high reliability in instrumentation which is subjected to vibrational environments. Alternatively, instead of realizing the solution of a differential equation as in all analogue filters, the functional requirement may be reformulated as a difference equation. This allows use of a finite data-sampling rate for implementation by a digital filter[3] whose relative viability is discussed at the end of Chapter 9.

1.1 The influence of advancing technology

Though the invention of the active filter followed close upon the introduction of the feedback amplifier, early designs[4] necessarily relied upon the thermionic valve or tube for activation. Such an approach has obvious limitations and its failure to compete commercially with passive filters is not surprising.

Not until the advent of the semiconductor transistor in the 1940's did the production of active filter networks become feasible. Initially, attention was focused on the realization of novel devices such as the negative-impedance converter and their employment in design philosophies[5] firmly based on the use of discrete components. However, like the transistor amplifier, these devices required individual circuit design and the possibility of low-cost mass production did not arise until the development of the silicon monolithic operational amplifier in the mid-1960's. The impact of this small, cheap, and reliable device on the practical feasibility of active filters cannot be overstated. For the first time, it became conceivable that they would compete with passive filters in terms of price as well as performance. Indeed, good-quality active filters can be readily designed and built using the resistors, capacitors, and operational amplifiers stocked by most electronics laboratories.

Intense research effort on the development of minimized-sensitivity filter structures together with further technological advances (enabling improved operational amplifiers acting in conjunction with thin- and thick-film passive components trimmed to precision by computer-controlled laser beams) has led to increasing use of active filters. The high volume of commercial sales has been such that the USA active-filter market was already exceeding $15M in 1977. Yet, this figure takes no account of the in-house and custom-design capabilities of such corporations as Beckman Industries, Bell Laboratories, and General Instruments which are currently supplying vast quantities of filters for use in both public and private telephone systems.

1.2 Applications of active filters

The trepidation which many electronic engineers have for active filters arises from adverse publicity given to the problems of sensitivity associated with component variations. However, these problems are only significant when attempting to realize the strict specifications required for some communications applications and not, say, when designing a simple instrumentation filter. There can be little doubt that the rapid growth in large-scale telecommunication systems provides a major market impetus in active filters. Perhaps the most important use is in conjunction with PCM-CODEC (pulse-code modulation coder/decoder) chips for the digitization of telephones. This requires very tight filter specifications in terms of sharpness of cut-off and passband ripple. Further applications in communications include[6] de-emphasis

INTRODUCTION

and pre-emphasis, aliasing, equalization, active impedance-matching networks for repeaters, and teleprinter lowpass filters to suppress harmonics.

An interesting example of communications applications is in dual-tone multi-frequency (DTMF) signalling as developed by Bell Telephone Laboratories for use in Touch Tone† dialling and destined to become standard in the American telephone market. It is also used in associated systems such as the radiotelephone link on the Metroliner railway operating between New York City and Washington, DC. Each digit in the telephone number being called generates a

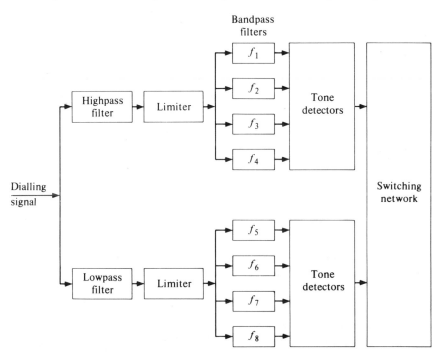

Fig. 1.1 Block diagram of Touch Tone receiver

unique pair of frequencies (one high-band and one low-band) which are subsequently detected by the decoder at a remote exchange. As illustrated in Fig. 1.1, the incoming signal is first divided into its high- and low-frequency components, limited, and then selected by a bank of bandpass filters. Tone detectors are used to convert the signal to d.c. form for appropriate control of the switching network.

The other major application of active filters lies in instrumentation, examples of which are: noise rejection in digital voltmeters, harmonic filtering in transformers, vibration, and shock studies. Not least of these is their increasing use in the field of medical electronics,[7] where they have a distinct advantage over passive filters in the low and sub-audio range. Many physiological signals

† Touch Tone is a registered trademark of the American Telephone & Telegraph Company.

fall within the three-decade band below 20 Hz, for example: heart monitoring by electrocardiograph (ECG) and brain-wave recording by electro-encephalograph (EEG). As well as providing size advantages in such applications, active filters overcome the problems of impedance matching inherent in passive systems. Although the low amplitude level of the signal means that noise introduced by the filters themselves is a problem, its effect can be reduced by providing low-noise amplification before the filtering stage.

Further areas of application are legion and include control (e.g., servo-system design and process-control equipment), entertainment electronics (e.g., stereo amplifiers and musical synthesis), and pulse shaping (e.g., to avoid crosstalk in colour TV broadcasting equipment). That is without mentioning sonar systems, telemetry, phase-locked loops, oscillators, hum suppression, pre-filtering of analogue-to-digital converters, spectrum analysis, and seismological research. Nor is this list comprehensive and areas of application not yet visualized will emerge as the twin effects of lower cost and greater production volumes create an increased awareness of active filters among engineers and other potential users.

1.3 The need for the text

The understanding and design of active filters inevitably necessitates the availability of an ordered and concise literature source emphasizing engineering perspective in the bewildering array of techniques rather than historical precedence. Recent books on the subject have largely been aimed at the specialist, either in the form of collected papers or as advanced reference manuals. As such, they are either unsuitable as introductory volumes or are quite beyond the pocket of the majority of students. Furthermore, the best of the introductory texts were written several years ago and of necessity lack in perspective as well as in discussion of the more recent developments consequent upon advances in microelectronics. The implications of new production techniques such as laser trimming clearly emphasize the need for an up-to-date book of moderate length which is suitable for introducing technically versed readers to this exciting and rapidly advancing field.

The treatment presupposes the basic knowledge normally covered prior to the final year of a degree course in electrical and/or electronic engineering. Accordingly, readers are expected to be familiar with the rudiments of circuit theory, including mesh/nodal analysis and two-port network notation, as well as to have an understanding of transistor circuits and device modelling. Nor does space permit the exposition of matrix analysis, Cramer's rule, or Gaussian elimination, despite the fact that familiarity with matrix operations as exemplified by the inverse and transpose are necessary in some sections of the book. However, a matrix conversion table is provided in the appendices for ready reference while the serious reader will overcome areas of unfamiliarity in the knowledge that the mathematics used is of the truly essential nature.

1.4 Outline of contents

After the general introduction of the present chapter, attention is directed to consideration of the approximation problem in Chapter 2. Although little more than a summary of standard approximating functions, the material serves to create a general awareness of the problem while furnishing sufficient data to supply the design examples of later chapters. Perhaps to an even greater extent, the discussion of passive network synthesis in Chapter 3 is a concession to realism rather than theoretical fantasy. The chapter surveys the most relevant properties of *RC* networks, culminating in a tabulation of results for subsequent use in design situations.

Most significant, however, is the energy foundation for realizability which forges a unique link between this topic and the discussion of selectivity and sensitivity in the succeeding chapter. The three concepts are thus unified in terms of energy—but only on the basis of one-port theory as no solution exists yet for the corresponding two-port analysis. Turning to Chapter 4, the topics of selectivity, sensitivity, and stability are inextricably related in filter network studies. A realistic approach to one necessitates an understanding of the others. Selectivity must be considered first since it prescribes response and can be the sole aim of design. Its enhancement requires the introduction of active elements in arrangements which are inherently associated with sensitivity and stability problems. After presenting the classical measures of sensitivity, the relative importance of frequency sensitivity as compared with Q sensitivity for second-order sections is highlighted before the reader is introduced to the latest multi-parameter statistical concepts.

The need for an active element to enhance selectivity leads naturally to Chapter 5, in which specific devices and their models are considered. On account of its ready availability in integrated form, particular emphasis is placed on the role of the operational amplifier as a filter element in its own right and also as the building block for other devices such as gyrators and generalized impedance converters. The amplifier is modelled using the nodal admittance matrix, which lends itself to ready incorporation in computer-based analysis and allows straightforward introduction of non-ideal parameters in later chapters. The material is supplemented by an appendix on the construction and ordered reduction of the nodal admittance matrix.

Formal active-network realization techniques are introduced in Chapter 6 through the medium of the controlled source. Various possibilities are investigated including single, dual, and mixed-source realizations. As in the succeeding chapters, the customary 'short-cut' approach of coefficient matching is usually adopted for the reasons stated in Chapter 3. Networks containing a single operational amplifier are featured in Chapter 7, where the active element is used in both its single-ended and differential-input forms. Exploitation of the latter to afford greater flexibility together with selectivity improvement by controlled positive feedback is described. The analogue techniques of Chapter 8 include the ladder-based inductor-replacement methods using gyrators and generalized impedance converters as well as the analogue-computer-derived methods such as the renowned leapfrog and state-variable realizations. Interest in these multi-amplifier multiloop feedback techniques has increased in recent years with the inexpensive availability of integrated operational amplifiers and is encouraged by the generally low sensitivity properties of the ladder system.

Overall comparisons of these various approaches on the basis of second-order bandpass realizations are made in Chapter 9 and the importance of the particular choice of design conditions is emphasized. This leads to discussion of the all-important trade-off between high-frequency performance and substrate requirements as manifested by component spread. The merits of the popular gain-sensitivity product as an active-filter performance comparison criterion are considered and its emergence as a degenerate case of the multiparameter measure demonstrated before a more comprehensive figure of merit is suggested. But the main purpose of Chapter 9 is to provide practical perspective for the theoretical concepts of the earlier text. Current fabrication technologies are reviewed and the manifold effects of component imperfections described together with techniques adopted for minimizing their influence. The importance of amplifier gain-bandwidth product for higher-frequency selective filters is stressed and a preference is stated for certain network configurations in these circumstances. Of the technological properties tabulated, environmental susceptibility as exemplified by temperature coefficient is among the more important and means of reducing the effect of drift on performance are indicated. After introducing modern automated tuning procedures, the implementation of high-order filter systems is discussed before a section is devoted to commercially available filters to reflect the extent of market activity and the types of filter forming standard product lines. Finally, the implications of expected future developments are considered from the point of view of the viability of active filters in the face of competition from digital signal-processing systems.

Attention is drawn to the *sample design* of Appendix 6, which is intended to give clear advice to those readers desirous of directly implementing a practical filter. Starting from a typical response specification, the procedure is carried through in its entirety to yield physically realizable component values.

INTRODUCTION

1.5 Special features

The treatment of the subject is characterized by some distinctive features, notably the reliance on the design tables in Chapters 6, 7, and 8, which thence act as pillars of comparison. Also, judicious use is made of genealogical perspective charts at the end of chapters to provide a rapid initial guide to the various techniques and realization procedures. Worked numerical examples are sprinkled through the chapters, although, in keeping with a text of this level, no problems are included for solution by the reader. The temptation to incorporate specific computer programs has been resisted since, to be meaningful, they would occupy several pages of text. Furthermore, the existence of programmable pocket calculators often facilitates routine evaluations without recourse to large machines. Network analysis and synthesis programs for use in conjunction with the latter are already[8] well documented.

An attempt has been made throughout to adhere to standard symbology; however, occasional deviations from accepted notation have been necessary to maintain consistency within the text. Such modifications are kept to a minimum and, in cases where several identifiers exist, the consensus view is adopted. For ease of recognition in matrix arrays and equations, the admittance parameters of passive networks are denoted by y_{ij} while those of active networks by Y_{ij}. Single-suffix notation is used to characterize discrete circuit elements as y_i and Y_i. Also, since it is sometimes convenient to analyse in terms of admittances and on other occasions in terms of impedances, symbols such as Y_i and Z_i are taken to represent the same element. This is particularly so in the case of resistances R_i and conductances G_i which are used with inverted interchangeability as is opportune. For clarification of diagrams, a general immittance is symbolized by a rectangle while specific conductance or resistance is shown in the traditional zigzag form.

Other features to note are the use of 'grounded' rather than 'earthed' for consistency with the majority of networks literature and the usual adoption of equal-valued-capacitor designs for simplicity. When discussing selectivity levels, 'medium Q' is taken as the range $10 \to 50$ while 'high Q' and 'low Q' are the ranges above and below this, respectively.

As intimated in Sec. 1.3, the major appeal of this book is expected to be in universities and other institutes of higher education where it is hoped to fulfill a need for a relatively concise text suitable for final-year undergraduate or first-year postgraduate courses. It constitutes a complete course in active filters which can be used to complement lectures on passive network synthesis. Alternatively, it can be used as part of broader options in communications, instrumentation, and control systems. The material should also prove suitable as an initial literature guide and reference text for postgraduate research students. Of the many anticipated roles of the book, not least is that of serving the requirements of practical engineers in industry who are seeking a rapid entry into the field of active filters. This text is an attempt to meet these various demands by achieving the realistic approach necessary in practical situations without sacrificing the rigour essential to the study of active networks.

References

1. *IEEE Standard Dictionary of Electrical and Electronic Terms*, Wiley Interscience, New York, 1972.
2. G. C. Temes and S. K. Mitra, *Modern Filter Theory and Design*, Wiley, New York and Chichester, 1973.
3. L. R. Rabiner and B. Gold, *Theory and Applications of Data Sample Processing*, Prentice Hall, Englewood Cliffs, N.J., 1975.
4. H. H. Scott, 'A new type of selective circuit and some applications', *Proc. IRE*, **26** (2), 226–235, 1938.
5. J. G. Linvill, 'Transistor negative-impedance converters', *Proc. IRE*, **41**, 725–729, 1953.
6. F. G. Stremler, *Introduction to Communications Systems*, Addison Wesley, Reading, Mass., 1977.
7. F. Ellern, 'Active physiologic signals', *Medical Electronics and Data*, **6**, 36–39, Jan.–Feb. 1975.
8. D. Calahan, *Computer-aided Network Design*, McGraw-Hill, New York, 1972.

2. The approximation problem

Filter performance is customarily prescribed graphically in the frequency or time domain. The former is more usual, being characterized by amplitude and phase-response plots against frequency. However, since electrical-network reactances have continuous frequency characteristics (except at a resonance),

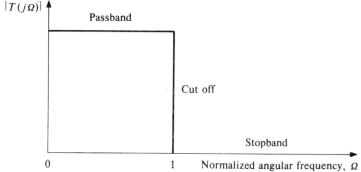

Fig. 2.1. Ideal lowpass magnitude characteristic

then the abrupt cut-off inherent in the ideal response of Fig. 2.1 cannot actually be attained by any finite connection of elements. Hence arises the so-called 'approximation problem': that is, the determination of system functions which approximate the given curve within specified tolerances and which are at the same time realizable as physical networks. Alternatively, computational techniques based on least-squares[1] methods, or interpolation at a set of points in a frequency band, may be applied to generate functions and element values which optimize performance in a given situation.

Exact synthesis[2] of filter networks requires that the approximating functions be expressed as the quotient of polynomials in the complex-frequency variable $s = \sigma + j\omega$, where the imaginary part ω is the angular frequency. Since excitation is normally a steady-state sinusoidal signal, then it is only the behaviour along the $s = j\omega$ axis which is considered here. For a given specification, it is necessary to find the lowest-order rational transfer function

$$T(s) = \frac{N(s)}{D(s)} = \frac{a_m s^m + a_{m-1} s^{m-1} + \cdots + a_2 s^2 + a_1 s + a_0}{b_n s^n + b_{n-1} s^{n-1} + \cdots + b_2 s^2 + b_1 s + b_0} = \frac{\sum_{i=0}^{m} a_i s^i}{\sum_{i=0}^{n} b_i s^i} \quad (2.1)$$

which must satisfy stability conditions and have $m \leqslant n$ in order to ensure realizability as a lumped-linear network.

$T(s)$ is usually a minimum-phase function (no zeros in the right half-plane), but more demanding specifications often necessitate use of non-minimum phase. The approximation process consists of determining the particular values of the real coefficients a_i and b_i as a prelude to synthesis. Evaluation of pole and zero positions allows expression in factored form which is suitable for cascade realization. There is no unique solution for the realization of a response to meet the usual forms of specification and a variety of standard functions are used for convenience. A range of such functions is now described, each offering particular features with regard to magnitude and/or phase approximation.

2.1 Magnitude approximations

The ideal characteristic in Fig. 2.1 is the basic lowpass response passing low-frequency signals and rejecting high-frequency signals above the normalized angular frequency $\Omega = (\omega/\omega_0)$ having cut-off value $\Omega_0 = 1$. By the reasoning above, the sharp discontinuity can only be approximated to within a stipulated error giving rise to the typical practical bounds on the response, as indicated in Fig. 2.2. Attention is initially focused on the lowpass prototype with transmission zeros at infinity (i.e., $a_i|_{i \neq 0} = 0$ or $N(s) = 1$) for which several solutions emerge from the general transfer-function amplitude

$$\gamma = 20 \log_{10} \frac{|T(j\Omega)|}{T(0)} = 10 \log_{10} \left[\frac{1}{1 + f(\Omega^2)} \right] \quad \text{dB} \quad (2.2)$$

where $f(\Omega^2)$ is the characterizing function, γ_{max} is the maximum allowable loss at ω_1, and γ_{min} the minimum acceptable attenuation at ω_2.

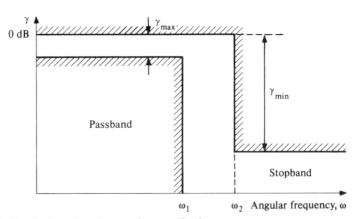

Fig. 2.2. Practical bounds on lowpass filter specifications

THE APPROXIMATION PROBLEM

2.1.1 Monotonic approximations

Monotonicity of response is most readily assured by requiring maximally flat behaviour at zero frequency. Discarding the decibels in Eq. (2.2) and applying a binomial approximation gives the error function around zero frequency as

$$\frac{T(j\Omega) - T(0)}{T(0)} \simeq -\tfrac{1}{2} f(\Omega^2)$$

In order that this has the maximum number of derivatives equal to zero at the frequency origin, it is necessary to choose $f(\Omega^2) = B_n(\Omega) = \varepsilon^2 \Omega^{2n}$, where

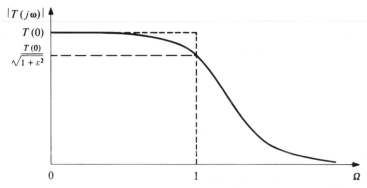

Fig. 2.3. Butterworth approximation response

$B_n(\Omega)$ is the Butterworth polynomial and ε is a real constant. Termed the *maximally flat* or *Butterworth*[3] approximation, the corresponding magnitude function is

$$|T(j\Omega)| = \frac{T(0)}{1 + \varepsilon^2 \Omega^{2n}} \qquad (2.3)$$

which, as illustrated in Fig. 2.3, decreases monotonically to a loss of $10 \log_{10}(1 + \varepsilon^2)$ dB at $\Omega = 1$ and subsequently to infinite attenuation at $\Omega = \infty$. Thus, for a specified band-edge loss of γ_{max}, the required constant is determined by

$$\varepsilon = \sqrt{10^{\gamma_{max}/10} - 1} \qquad (2.4)$$

In the normalized form, $\varepsilon = 1$ and the response level is 3 dB down at $\Omega = 1$.

The Laplace transfer function can be derived by using the concept of 'analytic continuation',[2] the poles being given by the left-half-plane roots of

$$D(s)D(-s) = 1 + (-s^2)^n$$

where s is now taken as $j\Omega$ for convenience in the remainder of the approximation discussion only. These are located on a semi-circle in the

Table 2.1 Factored-form coefficients for various approximating functions

$$D(s) = \begin{cases} \prod_{i=1}^{n/2} (s^2 + b_{1i}s + b_{0i}), & n = 2, 4, 6 \\ (s + b_1) \prod_{i=1}^{(n-1)/2} (s^2 + b_{1i}s + b_{0i}), & n = 3, 5 \end{cases}$$

Approximation	Order n	b_1	b_{11}	b_{01}	b_{12}	b_{02}	b_{13}	b_{03}
Maximally flat magnitude (Butterworth)	2	—	1.414 214	1	—	—	—	—
	3	1	1	1	—	—	—	—
	4	—	0.765 367	1	1.847 759	1	—	—
	5	1	0.618 034	1	1.618 034	1	—	—
	6	—	0.517 638	1	1.414 214	1	1.931 852	1
0.5 dB passband equi-ripple (Chebyshev)	2	—	1.425 624	1.516 203	—	—	—	—
	3	0.626 457	0.626 456	1.142 448	—	—	—	—
	4	—	0.350 706	1.063 519	0.846 680	0.356 412	—	—
	5	0.362 320	0.223 926	1.035 784	0.586 245	0.476 767	0.579 588	—
	6	—	0.155 300	1.023 023	0.424 288	0.590 136	0.579 588	0.156 997

Factored-form coefficients of $D(s)$

THE APPROXIMATION PROBLEM

	n							
1 dB passband equi-ripple (Chebyshev)	2	—	1.097 734	1.102 510	—	—	—	—
	3	0.494 171	0.494 171	0.994 205	—	—	—	—
	4	—	0.279 072	0.986 505	0.673 739	0.279 398	—	—
	5	0.289 493	0.178 917	0.988 315	0.468 410	0.429 298	—	—
	6	—	0.124 362	0.990 733	0.339 763	0.557 720	0.464 125	0.124 707
Maximally flat delay (Thomson)	2	—	3	3	—	—	—	—
	3	2.322 185	3.677 814	6.459 432	—	—	—	—
	4	—	5.792 422	9.140 133	4.207 578	11.487 799	—	—
	5	3.646 739	6.703 912	14.272 476	4.649 348	18.156 314	—	—
	6	—	8.496 718	18.801 128	7.471 416	20.852 819	5.031 864	26.514 025
Gaussian	2	—	2.579 265	1.948 503	—	—	—	—
	3	1.520 050	2.765 142	2.806 807	—	—	—	—
	4	—	2.842 111	3.647 126	3.261 485	2.815 016	—	—
	5	1.776 725	2.896 501	4.542 015	3.409 013	3.422 784	—	—
	6	—	2.937 543	5.479 213	3.516 602	4.096 024	3.754 237	3.628 079

complex plane with positions evaluated from

$$s_k = -\sin\left(\frac{2k-1}{n}\right)\frac{\pi}{2} + j\cos\left(\frac{2k-1}{n}\right)\frac{\pi}{2} \quad (2.5)$$

with $k = 1, 2, \ldots, n$. Denominator polynomial $D(s)$ has coefficients given by the recursive relation

$$b_k = \frac{\cos\left(\frac{k-1}{n}\frac{\pi}{2}\right)}{\sin\left(\frac{k}{n}\frac{\pi}{2}\right)} b_{k-1} \quad (2.6)$$

with $b_0 = 1$, these being listed in factored form for several values of n in Table 2.1. Manipulation[4] of Eq. (2.4) shows that the minimum order of Butterworth approximation to meet the filter specification of Fig. 2.2 is

$$n_B = \frac{1}{2} \cdot \frac{\log_{10}[(10^{v_{\min}/10} - 1)/(10^{v_{\max}/10} - 1)]}{\log_{10}(\omega_2/\omega_1)} \quad (2.7)$$

where ω is now adopted for specifications. A sharp cut-off characteristic with $\omega_1 \to \omega_2$ obviously requires a high-order filter. For a more rapid transition from pass to stopband, *optimum 'L'* approximations[5] can be made by choosing $f(\Omega^2) = L_n(\Omega)$ as Legendre polynomials.

2.1.2 Equi-ripple approximations

To achieve a significantly sharper transition, it is necessary to relax the previous monotonic restraint and spread the error over a finite frequency band. Instead of matching the ideal response only at $\Omega = 0$, the equi-ripple approximation is equally good at $\Omega = 1$ by allowing the characteristic to ripple between a series of maxima and minima. This is known[6] to be the optimum solution to the problem of minimizing rational-function order for the smallest amplitude deviation over a finite frequency band.

Chebyshev approximation The equi-ripple requirement can be met in the passband by choosing the characterizing function in Eq. (2.2) such that it ripples between 0 and 1 in the $\Omega = 0 \to 1$ band while becoming very large in the stopband. The appropriate characterizing function in the range $0 \leqslant \Omega \leqslant 1$ is

$$f(\Omega^2) = \varepsilon^2 C_n^2(\Omega) = \varepsilon^2 \cos^2[n \cos^{-1}(\Omega)] \quad (2.8)$$

where ε is again a real constant <1 and a recursive formula can be used to evaluate the Chebyshev[7] polynomial,

$$C_n(\Omega) = 2\Omega C_{n-1}(\Omega) - C_{n-2}(\Omega) \quad (2.9)$$

THE APPROXIMATION PROBLEM

with $C_0(\Omega) = 1$ and $C_1(\Omega) = \Omega$. The resulting type of amplitude response is illustrated in Fig. 2.4 for an even function. If n is odd then the ripple starts from the $T(0)$ level. In both cases, the transfer gain at cut-off is

$$T(\Omega_0) = \frac{T(0)}{\sqrt{1+\varepsilon^2}}$$

so that the permissible ripple determines the value of ε. The number of half-cycles of ripple in the passband is equal to the order n of the function which is,

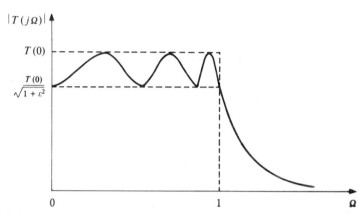

Fig. 2.4. Sixth-order Chebyshev approximation response

in turn, dictated by the sharpness of cut-off required outside the passband. When $\Omega > 1$, it is more convenient to express the Chebyshev polynomial as

$$C_n(\Omega) = \cosh[n \cosh^{-1}(\Omega)] \qquad (2.10)$$

Then, by Eqs. (2.2), (2.8), and (2.10), the minimum order of Chebyshev approximation[4] to satisfy the bounds of Fig. 2.2 is

$$n_C = \frac{\cosh^{-1}[(10^{p_{min}/10} - 1)/(10^{p_{max}/10} - 1)]^{1/2}}{\cosh^{-1}(\omega_2/\omega_1)} \qquad (2.11)$$

whose nature ($n_C < n_B$) permits a simpler filter than the Butterworth for identical performance specification.

Application of analytic continuation yields the left-half-plane pole locations of the Chebyshev function as

$$s_k = -\sinh a \cdot \sin\left(\frac{2k-1}{n}\right)\frac{\pi}{2} + j \cosh a \cdot \cos\left(\frac{2k-1}{n}\right)\frac{\pi}{2} \qquad (2.12)$$

with $k = 0, 1, 2, \ldots, (2n-1)$ and $\quad a = \frac{1}{n} \sinh^{-1}\left(\frac{1}{\varepsilon}\right)$

Squaring and adding real and imaginary parts readily shows these to lie on an ellipse in the complex plane. Normalizing the major axis with respect to cosh a allows determination of Chebyshev poles from Butterworth by a tanh a scaling (see Fig. 2.5). Polynomial factors for several orders of $D(s)$, having passband ripples of 0.5 dB and 1 dB, are presented in Table 2.1.

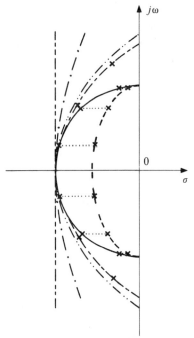

Fig. 2.5. Pole distributions in complex plane for different approximations. ———, Butterworth; —··—, catenary; -----, Chebyshev; ————, linear phase; ———, parabolic; —·—, Thomson

Inverse Chebyshev approximation[2,4] If the absolute power level of the response in Fig. 2.4 is subtracted from the upper limit and Ω is replaced by $(1/\Omega)$, then a response function

$$\gamma = 10 \log_{10} \left(\frac{1}{1 + 1/\varepsilon^2 C_n^2(1/\Omega)} \right) \quad \text{dB} \quad (2.13)$$

emerges which is maximally flat in the passband and equi-ripple in the stopband. However, such a function contains finite transmission zeros and its synthesis is consequently more involved.

Elliptic approximation It might be expected that the response which allows greatest flexibility is that which exhibits equi-ripple behaviour in both the passband and the stopband, as in Fig. 2.6. The characterizing function

$$f(\Omega^2) = \varepsilon^2 R^2(\Omega) \quad (2.14)$$

where

$$R^2(\Omega) = \frac{\prod_{i=1}^{n/2}(1 - a_{0i}\Omega^2)}{\prod_{i=0}^{n/2}(a_{0i} - \Omega^2)} \quad (2.15)$$

if even, and an extra Ω factor if odd, meets the associated requirements of oscillating between maxima and minima in two bands but implies conditions on

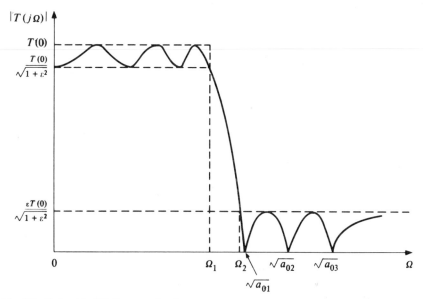

Fig. 2.6. Sixth-order elliptic approximation response

the function differential. From this devolve the elliptic integral functions[6,7,8] whose roots can be found by trigonometric transformation. Network synthesis then necessitates numerical evaluation by computer-aided techniques or resort to tabulated[9] data.

Elliptic functions themselves are well documented elsewhere[4] and, for the purposes of the present text, it is sufficient to approximate the transfer function by the multiple notch

$$T(s) = \prod_{i=1}^{n/2} \frac{H(s^2 + a_{0i})}{(s^2 + b_{1i}s + b_{0i})} \quad (2.16)$$

where $H = b_{0i}/a_{0i}$. This is the even-order formulation, the transmission zeros $j\sqrt{a_{0i}}$ representing the frequencies of infinite loss. The odd-order function has a similar form but with the addition of a real pole which ensures zero instead of finite transmission at infinite frequency. A selection of coefficients for the

Table 2.2 Coefficients and design parameters for elliptic approximation

(a) Second-order lowpass elliptic approximation

θ (degrees)	Ω_2/Ω_1	a_{01}	1 dB ripple			2 dB ripple			3 dB ripple		
			b_{11}	b_{01}	A_{min} (dB)	b_{11}	b_{01}	A_{min} (dB)	b_{11}	b_{01}	A_{min} (dB)
20	2.92	16.582	1.0577	1.1347	23.3	0.7822	0.8480	25.8	0.6288	0.7287	27.1
25	2.37	10.673	1.0324	1.1513	19.3	0.7687	0.8617	21.7	0.6196	0.7410	23.2
30	2.00	7.464	0.9989	1.1701	16.1	0.7508	0.8783	18.6	0.6075	0.7559	19.9
35	1.74	5.5295	0.9559	1.1896	13.3	0.7278	0.8973	15.8	0.5919	0.7736	17.1
40	1.56	4.2743	0.9019	1.2083	11.0	0.6988	0.9184	13.4	0.5721	0.7938	14.6
45	1.41	3.4142	0.8360	1.2240	8.9	0.6627	0.9409	11.2	0.5474	0.8165	12.4
50	1.31	2.7995	0.7580	1.2343	7.1	0.6185	0.9638	9.3	0.5169	0.8412	10.3

(b) Fourth-order lowpass elliptic approximation with 3 dB passband ripple

θ (degrees)	Ω_2/Ω_1	a_{01}	a_{02}	b_{11}	b_{01}	b_{12}	b_{02}	A_{min} (dB)
20	2.92	55.385	9.9264	0.4178	0.2068	0.1613	0.9090	69.32
25	2.37	35.20	6.4688	0.4219	0.2134	0.1563	0.9125	61.36
30	2.00	24.227	4.5933	0.4271	0.2218	0.1502	0.9167	54.76
35	1.74	17.603	3.4653	0.4332	0.2324	0.1430	0.9216	49.09
40	1.56	13.294	2.7364	0.4404	0.2456	0.1346	0.9273	44.07
45	1.41	10.330	2.2401	0.4485	0.2618	0.1250	0.9336	39.52
50	1.31	8.199	1.8888	0.4576	0.2819	0.1142	0.9407	35.33

second- and fourth-order functions is presented in Table 2.2. The characteristic provides a more rapid transition from passband to stopband than does the Chebyshev (and hence Butterworth) form, but at the expense of a finite attenuation level in the stopband. The defining parameters are

(a) the order of the function (n)
(b) the frequency transition ratio (Ω_2/Ω_1) or modular angle $\theta = \sin(\Omega_1/\Omega_2)$
(c) the level of passband and/or stopband ripple (both defined by ε)

Remember that only (a) is required for a normalized ($\varepsilon = 1$) Butterworth approximation; (a) and (b) for a Chebyshev.

2.2 Delay and related approximations

The magnitude approximations discussed above take no account of the associated phase shift. On account of the relative insensitivity of the human ear to phase distortion, this is justifiable for aural communication. However, it is a different matter in digital transmission systems since the phase characteristic affects such pulse properties as rise time and overshoot. For distortionless transmission through a network, the output signal V_2 should replicate the input signal V_1, though it may be changed in amplitude and delayed by a constant time τ such that

$$V_2(t) = V_1(t - \tau)$$

Taking the Laplace transform gives the transfer function of a pure delay as

$$T(s) = \frac{V_2(s)}{V_1(s)} = e^{-s\tau} \qquad (2.17)$$

For convenience, the quality of a phase response is measured by the group delay

$$\Delta(\Omega) = -\frac{d}{d\Omega}[\arg T(j\Omega)] \qquad (2.18)$$

For the linear phase shift function of Eq. (2.17), this is constant, thereby ensuring undistorted signal transmission. However, in practice, this can only be realized by use of lossless delay lines. Implementation with lumped network elements necessarily gives rise to phase non-linearity which delays different Fourier components by different amounts, resulting in distortion. Consequently, constant $\Delta(\Omega)$ can only be approximated over a finite frequency band by developing a rational representation prior to synthesis.

2.2.1 The maximally flat delay

The most obvious approximation is that resulting from Maclaurin expansion of Eq. (2.17), giving $T(s) = H/B_n(s)$, where the recursive relation for the Bessel polynomial is

$$B_n(s) = (2n - 1)B_{n-1}(s) + s^2 B_{n-2}(s) \qquad (2.19)$$

with $B_0 = 1$ and $B_1 = s + 1$. In fact, it approximates the ideal group delay in a maximally flat manner. Polynomial factors for several orders in the Thomson approximation[10] are again listed in Table 2.1, while pole locations to the order of 31 are available.[11]

2.2.2 Other approximations for pulse applications

While Thomson filters have negligible overshoot in their transient response, their rise times are longer than for the corresponding Butterworth filter. On the other hand, the latter has an undesirable level of overshoot which increases with order. A compromise is available[12] in the *transitional Butterworth–Thomson approximation*, which provides a trade-off between characteristics rather than simultaneous advantages. In the same manner, a *transitional Butterworth–Chebyshev approximation* can be developed[13] to trade-off amplitude characteristics.

Returning to the pole distributions of Fig. 2.5, linear phase response would appear to be associated with a pole distribution having more nearly the same real parts. In the limit, this is a pole distribution parallel to the $j\omega$ axis. Hence approximations with pole distributions between the Butterworth semi-circle and this vertical line should provide good transient-response properties. The *parabolic* and the *catenary* filter classes have been suggested and compared,[14] the terms describing the functional distributions in the complex plane of Fig. 2.5.

Of particular interest in pulse communications is the *Gaussian* characteristic, deriving its name from the shape of its impulse response. The step response, being the integral of the latter, is consequently free from overshoots. However, the ideal Gaussian is a theoretical abstraction since its true zero can only be found at negative-infinite time, implying infinite delay and non-realizability. It can be approximated for realization as a finite network by developing a Taylor-series polynomial[15] of the squared modulus

$$|T(j\Omega)|^2 = \exp(-\beta\Omega^2)$$

$$= 1 + \frac{2\beta}{1!}\Omega^2 + \frac{(2\beta)^2}{2!}\Omega^4 + \frac{(2\beta)^3}{3!}\Omega^6 + \cdots \qquad (2.20)$$

and then truncating after n terms. Analytic continuation techniques as in Sec. 2.1 then allow determination of the roots and factor coefficients entered in

Table 2.1 for a selection of polynomials. The resulting filter has a better step response than the Thomson filter but inferior magnitude response. Better performance for a given order can be obtained by expanding the Gaussian into the Laguerre polynomial[16] and taking account of missing terms after truncation.

Among other filter functions available is the *equi-ripple delay approximation*[17] to a linear phase characteristic, which may also be approached on a cut-and-try basis by assuming a rectilinear distribution of poles and adjusting the positions of the end two.

2.2.3 Delay equalization

The relatively poor magnitude response attendant to the phase approximations discussed above limits their application as delay equalizers. These have the requirements of constant amplitude and prescribed phase which are not easily achieved[6] simultaneously with all-pole functions. It is more realistic to supplement the pole pattern by a symmetrically placed linear array of right-half-plane zeros, thereby doubling the phase but maintaining constant amplitude, i.e., the *allpass response*. For the second-order case,

$$T(s) = \frac{(s-p)(s-p^*)}{(s+p)(s+p^*)} = \frac{s^2 - a_1 s + a_0}{s^2 + a_1 s + a_0} \quad (2.21)$$

2.3 Frequency transformations

Consideration of approximations has hitherto been restricted to the normalized lowpass functions for which extensive tabulated data are available. When highpass, bandpass, or bandstop filters are prescribed, as indicated in Fig. 2.7, their specifications are best transformed into equivalent lowpass parameters. Having then determined the most suitable approximating function, application of the appropriate frequency transformation yields a function meeting the desired performance criteria.

The highpass specifications of Fig. 2.7(a) can be converted into equivalent lowpass terms by application of the transformation at the passband edge

$$p = \frac{\omega_{2/s}}{s} \quad (2.22)$$

producing a lowpass function in the transformed variable p with a passband from d.c. to $\Omega = 1$ and a stopband from $\Omega = \Omega_2/\Omega_1$ to infinity. After satisfying this with a standard approximating function, the appropriate highpass

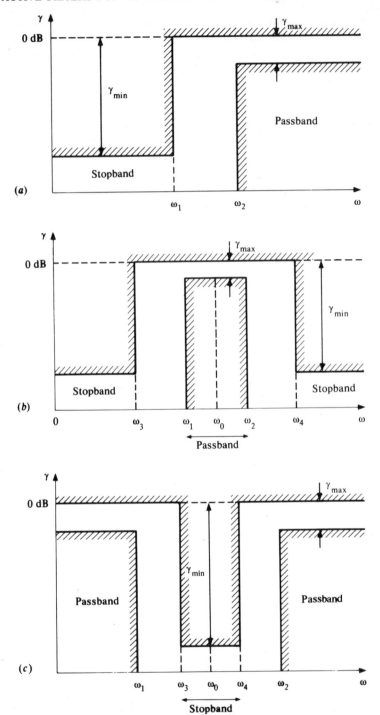

Fig. 2.7. Practical bounds on filter specifications. (a) Highpass; (b) bandpass; (c) bandstop

Table 2.3 Frequency transformations

Function	Frequency variable transformation	Specification transformation			
		Filter specification		Normalized lowpass	
		Passband	Stopband	Passband	Stopband
Highpass	$p = \dfrac{\omega_2}{s}$	$\omega_2 \to \infty$	$0 \to \omega_1$	$0 \to 1$	$\left(\dfrac{\omega_2}{\omega_1}\right) \to \infty$
Bandpass	$p = Q\left(\dfrac{s}{\omega_0} + \dfrac{\omega_0}{s}\right)$	$\omega_1 \to \omega_2$	$0 \to \omega_3$ $\omega_4 \to \infty$	$0 \to 1$	$\dfrac{\omega_4 - \omega_3}{\omega_2 - \omega_1} \to \infty$
Bandstop	$p = \dfrac{1}{Q}\,\dfrac{1}{\dfrac{s}{\omega_0} + \dfrac{\omega_0}{s}}$	$0 \to \omega_1$ $\omega_2 \to \infty$	$\omega_3 \to \omega_4$	$0 \to 1$	$\dfrac{\omega_2 - \omega_1}{\omega_4 - \omega_3} \to \infty$

characteristic emerges by application of Eq. (2.22). In the case of the symmetrical bandpass specifications ($\omega_1\omega_2 = \omega_3\omega_4 = \omega_0^2$) of Fig. 2.7(b), the required transformation to normalized lowpass is

$$p = Q\left(\frac{s}{\omega_0} + \frac{\omega_0}{s}\right) \qquad (2.23)$$

for which, solution at Ω_1 ($=\omega_1/\omega_0$) and Ω_2 ($=\omega_2/\omega_0$) yields

$$\Omega_1\Omega_2 = 1 \quad \text{and} \quad Q = \left(\frac{1}{\Omega_2 - \Omega_1}\right)$$

The corresponding transformed lowpass then has a stopband from $\Omega = (\Omega_4 - \Omega_3)/(\Omega_2 - \Omega_1) \to \infty$. A similar procedure for the symmetrical bandstop response bounds of Fig. 2.7(c) requires transformation

$$p = \frac{1}{Q}\left(\frac{1}{\dfrac{s}{\omega_0} + \dfrac{\omega_0}{s}}\right) \qquad (2.24)$$

with transformed lowpass stopband $\Omega = (\Omega_2 - \Omega_1)/(\Omega_4 - \Omega_3) \to \infty$. The various frequency transformations are summarized in Table 2.3 and a numerical design example is developed in Appendix 6. Asymmetrical responses are considered elsewhere.[4]

The approximations discussed in this chapter are intended for synthesis by lumped networks. If realization is to be in terms of distributed networks (see Sec. 3.5) of identical electrical length, then it is necessary to transform the complex frequency variable to $t = \tanh s$. For digital-filter implementation (see Sec. 9.8), the transformation is $z = e^{-sT}$, where T is the sampling period.

References

1. F. F. Kuo, *Network Analysis and Synthesis*, Wiley, New York, 1966, Chap. 13, p. 366.
2. L. Weinberg, *Network Analysis and Synthesis*, McGraw-Hill, New York, 1962.
3. S. Butterworth, 'On the theory of filter amplifiers', *Exp. Wireless and Wireless Eng.*, **7**, 536–541, 1930.
4. R. W. Daniels, *Approximation Methods for Electronic Filter Design*, McGraw-Hill, New York, 1974.
5. A. Papoulis, 'Optimum filters with monotonic response', *Proc. IRE*, **46**, (3), 606–609, 1958.
6. J. D. Rhodes, *Theory of Electrical Filters*, Wiley, Chichester, 1976.
7. W. Cauer, *Synthesis of Linear Communication Networks*, McGraw-Hill, New York, 1958. (Based on work carried out in the 1930's.)

8. S. Darlington, 'Synthesis of reactance four-poles', *J. Math. Phys.*, **18** (4), 257–353, 1939.
9. E. Christian and E. Eisenman, *Filter Design Tables and Graphs*, Wiley, New York, 1966.
10. W. E. Thomson, 'Networks with maximally flat delay,' *Wireless Engr*, **29**, 255–263, 1952.
11. H. J. Orchard, 'The roots of maximally flat delay polynomials', *IEEE Trans.*, **CT-12**, 453–454, 1965.
12. Y. Peless and T. Murakami, 'Analysis and synthesis of transitional Butterworth–Thomson filters and bandpass amplifiers', *RCA Rev.*, 60–94, March 1957.
13. A. Budak and P. Aronhime, 'Transitional Butterworth–Chebyshev filter', *IEEE Trans.*, **CT-18**, 413–415, 1971.
14. M. S. Ghausi and M. Adamowicz, 'A new class of filters for pulse applications', *J. Franklin Inst.*, **282** (1), 20–30, 1966.
15. M. Dishal, 'Gaussian-response filter design', *Electron. Commun.*, **36** (1), 3–26, 1959.
16. N. B. Jones, 'A Laguerre series approximation to the ideal Gaussian filter', *Radio Electron. Engr*, **40** (3), 151–155, 1970.
17. A. I. Zverev, *Handbook of Filter Synthesis*, Wiley, New York, 1967.

3. Passive networks

Once the prescribed performance criteria of a filter have been approximated by a rational driving-point or transfer function using the techniques of the previous chapter, there remains the *realization problem* of implementing the function by a physical network. One's first resort is to the theory of lumped-linear-finite-passive-bilateral networks[1] (LLFPBN's), that is, realizations made up of a bounded number of basic discrete elements (resistors, capacitors, and inductors) which do not intrinsically require excitation with a power source. Such structures possess the fundamental property of *reciprocity*,[2,3] which implies that the relationship between signal excitation and measurement is unchanged when the points of excitation and response are interchanged. Whereas the *analysis* of electrical networks is a quite straightforward procedure, the converse process of *synthesis* is more involved, requiring the realization of a network function for a given excitation and response. Basically, the technique is to systematically decompose the prescribed function into constituent subfunctions which are recognizable as simple combinations of components. The resulting subnetworks are then interconnected in a manner appropriate to the nature of the function (e.g., in parallel for an admittance function).

For a full treatment of such methods, the reader is referred to the excellent existing texts[1,3,4,5] on passive network synthesis. As the purpose of the present volume is specifically to describe active filter networks, the coverage of passive networks in this chapter is necessarily abbreviated. The treatment presented here is slanted towards network configurations which are amenable to incorporation in or simulation by the active networks to be discussed in later chapters.

3.1 Properties of immittance functions

Networks containing only two kinds of passive element are special cases of the general *RLC* situation but are simpler to consider and, as shall be seen, often more appropriate in filter application. The three possible categories are presented in Table 3.1 with their corresponding impedance characteristics. *LC* networks, having no resistance, are often termed *lossless* or *reactance*

PASSIVE NETWORKS

Table 3.1 Frequency characteristics of two-element-kind passive networks

Series		Parallel	
Circuit	Frequency characteristic	Circuit	Frequency characteristic

networks. The simple series connection in the table has

$$\text{impedance} \quad Z_{LC}(s) = \frac{1}{s}\left(s^2 L + \frac{1}{C}\right) \tag{3.1}$$

and

$$\text{reactance} \quad X(\omega) = \frac{1}{\omega}\left(\omega^2 L - \frac{1}{C}\right)$$

Hence the reactance plot has a *pole* (i.e., $X(\omega) \to \infty$) when $\omega = 0$ and conjugate *zeros* (i.e., $X(\omega) = 0$) when $\omega = \pm 1/\sqrt{LC}$, both lying on the $j\omega$ axis of the complex frequency plane, as in Fig. 3.1. Also, note that the reactance slope

$$\frac{\partial X(\omega)}{\partial \omega} = L + \frac{1}{\omega^2 C}$$

is always positive, a fact which ensures[4] that poles and zeros of higher order $Z(s)$ must be simple (i.e., not repeated) and alternate on the $j\omega$ axis. These properties are also true for the susceptance of an LC admittance.

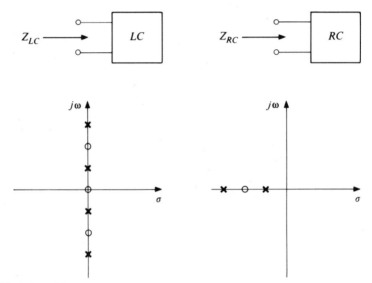

Fig. 3.1. Pole positions of one-port LC and RC networks

The RC series connection has impedance

$$Z_{RC}(s) = \frac{1}{s}\left(sR + \frac{1}{C}\right) \tag{3.2}$$

and, remembering the complex form of operator s, has a negative-real zero at $\sigma = -1/RC$. This time, the impedance slope $\partial Z_{RC}(\sigma)/\partial \sigma$ is always negative while the admittance slope is positive. In the parallel connection of Table 3.1, there is an analogy[1] between the familiar LC 'tank' resonance and that of the RC circuit which is 'resonant' in the sense that it passes no current when a voltage signal of type $e^{-t/RC}$ is applied rather than $e^{j\omega t}$. The properties of the LC function along the $j\omega$ axis hold for the RC function along the negative-real axis with the difference that the latter distinguishes between series impedances and parallel admittances.

PASSIVE NETWORKS

In the same manner, the properties of the RL circuit may be developed. Consider the admittance of the parallel connection

$$Y_{RL}(s) = \frac{1}{s}\left(sG + \frac{1}{L}\right) \tag{3.3}$$

which is recognizable as the same form as the series RC connection impedance function of Eq. (3.2) with conductance G instead of R and inductance L in place of C. In other words, an RL admittance function is identical with an RC impedance function and vice versa. This is confirmed by inspection of responses in Table 3.1 and consequently no separate study of RL immittances is necessary.

3.2 Transformations and duality

Frequency transformations from one type of filter response to another have been described in Table 2.3. Various network impedance transformations are now described which enable realization of alternative functions without the need for resynthesis and give rise to the general concept of duality.

3.2.1 LC:RC transformation

Comparing Eqs. (3.1) and (3.2), it is apparent that the impedance of an RC network can be determined from that of a topologically identical LC network by replacing inductance by resistance and s^2 by s, so that

$$Z_{RC}(s) = \frac{1}{\sqrt{s}} Z_{LC}(\sqrt{s}) \tag{3.4a}$$

or, more simply, by putting

$$L = R/s \tag{3.4b}$$

This effectively maps the $j\omega$-axis poles on to the negative-real axis of the complex plane (see Fig. 3.1). The corresponding reverse RC to LC transformation is described by

$$Z_{LC}(s) = sZ_{RC}(s^2) \tag{3.5a}$$

or just

$$R = sL \tag{3.5b}$$

Since the transfer ratio T of a network can be expressed as the ratio of immittance functions, then

$$T_{LC}(s) = T_{RC}(s^2) \tag{3.6}$$

By this means, the results of lumped LC synthesis may be readily applied to RC synthesis and immittance functions identified in the partitioning process to be described in Sec. 4.2.3.

3.2.2 RC:CR transformation

In two-element-kind realizations, it is often convenient to transform from lowpass to highpass filter functions by interchanging component types while preserving the topological graph. In the RC case, this involves:

(a) replacing each resistor R_i by a capacitor $C_i = 1/R_i$
(b) replacing each capacitor C_j by a resistor $R_j = 1/C_j$

This may be illustrated by reference to the simple section of Fig. 3.2(a), for which

$$Z_{RC}(s) = \frac{(R_1 + R_2) + sR_1R_2C_2}{1 + sR_2C_2} \tag{3.7}$$

whereas the transformed section in Fig. 3.2(b) has

$$Z_{CR}(s) = \frac{1 + sR_B(C_A + C_B)}{sC_A(1 + sR_BC_B)} = \frac{s(R_1 + R_2) + R_1R_2C_2}{s(s + R_2C_2)} \tag{3.8}$$

Inspection of Eqs. (3.7) and (3.8) shows that

$$Z_{CR}(s) = \frac{1}{s} Z_{RC}\left(\frac{1}{s}\right) \tag{3.9}$$

The result holds[6] for higher-order sections, for the transfer impedance of two-port networks, and for admittance characterization. By the ratio of functions like Eq. (3.9), the transfer ratio is transformed as

$$T_{CR}(s) = T_{RC}\left(\frac{1}{s}\right) \tag{3.10}$$

Significantly, these transformations are also valid when the network contains active components.

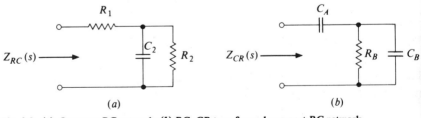

Fig. 3.2. (a) One-port RC network; (b) RC–CR transformed one-port RC network

PASSIVE NETWORKS

3.2.3 Duality and inversions

The principle of duality[7] expresses the correspondence between voltage and current differential equations arising from the mesh and nodal analyses of two related networks. Strictly, it is applicable only to networks with planar graphs (that is, those which can be drawn on a plane without any crossed branches). The procedure for setting up a dual configuration[1,3] of a given network can be explained by reference to the simple circuit in Fig. 3.3(a). First, a node is placed inside each mesh as well as an extra datum node outside the network. Then the nodes are linked by dashed lines through each element or source common to their meshes and the original elements replaced by their duals to form the network of Fig. 3.3(b).

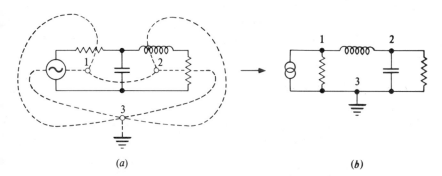

Fig. 3.3. (a) Construction of dual graph; (b) resistively inverted dual network

The nature of the elements in the dual network is determined[6] by the particular inversion employed. If $Z(s)$ and $Z_D(s)$ are generalized inverses of each other with respect to a real rational function $f(s)$, then

$$Z(s) \cdot Z_D(s) = f(s) \qquad (3.11)$$

The three simplest cases are:

(a) *Resistive inversion* with $f(s) = K$, a constant. It is illustrated by Fig. 3.3. However, in the context of RC-active networks, this is not particularly useful because the RC circuits are dualized into RL form.

(b) *Inductive inversion* with $f(s) = sK$, thereby transforming resistor R into inductor K/R and inductor L into resistor K/L. This may have application in gyrator realizations (see Chapter 8).

(c) *Capacitive inversion* with $f(s) = K/s$ converting capacitor C into resistor KC and resistor R into capacitor R/K. This RC dualization is perhaps the most useful in active-network considerations, often allowing transformation to more convenient circuit structures.

3.3 Realizability

The choice of approximation by the methods in Chapter 2 inherently assures that the resulting network function is realizable by a physical network. This implies that certain necessary conditions, mainly associated with stability considerations, are fulfilled. However, direct realization of the function is only possible for a *one-port network*, that is, with one pair of terminals defining both input and output variables. The network function is then either a *driving-point impedance* or *admittance* (jointly termed *immittance*), as illustrated in Fig. 3.1. Since, in practice, filters usually have separate pairs of terminals for input and output signals, then *the two-port network* (symbolized with terminal notation in Fig. 3.4) is of more concern. Here, the network function is described by a

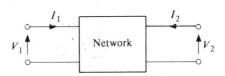

Fig. 3.4. Two-port network notation

transfer ratio of voltage (V_2/V_1), current (I_2/I_1), or immitance $(V_2/I_1, I_2/V_1)$ type from which a set of characterizing parameters must be derived to enable synthesis. The present section outlines procedures for ascertaining the realizability of driving-point and transfer functions in turn.

Although the driving-point and transfer functions of LLFPBN's take the form of quotients of rational polynomials in s, it does not follow that all such quotients automatically describe a physical network. Indeed, mathematically, there are a number of stringent requirements to meet which make it rather remarkable that any functions can be realized at all. This is best explained by resorting to energy principles. With the notation of Fig. 3.4, the total complex power delivered to the network is

$$[I^*][V] = [I^*]^t[R][I] + s[I^*]^t[L][I] + \frac{1}{s}[I^*]^t\left[\frac{1}{C}\right][I] \qquad (3.12)$$

where voltage column matrix

$$[V] = \begin{bmatrix} V_1 \\ V_2 \\ 0 \\ \vdots \\ 0 \end{bmatrix}$$

and transposed-conjugate current matrix

$$[I^*]^t = [I_1^* \, I_2^* \cdots I_n^*]$$

PASSIVE NETWORKS

the asterisk denoting complex conjugate and t the transpose. For n network loops, the resistance-parameter matrix is

$$[R] = \begin{bmatrix} R_{11} & R_{12} & \cdots & R_{1n} \\ R_{21} & R_{22} & \cdots & R_{2n} \\ \vdots & \vdots & & \vdots \\ R_{n1} & R_{n2} & \cdots & R_{nn} \end{bmatrix}$$

and likewise for the inductance and capacitance-parameter matrices $[L]$ and $[C]$, respectively. Equation (3.12) can be expressed as

$$[I^*]^t[V] = P_{d0} + sW_{m0} + \frac{W_{e0}}{s} \qquad (3.13)$$

where

$$P_{d0} = [I^*]^t[R][I], \quad W_{m0} = [I^*]^t[L][I], \quad \text{and} \quad W_{e0} = [I^*]^t\left[\frac{1}{C}\right][I]$$

are 'energy' functions.

3.3.1 Driving-point immittance functions

In the case of a single pair of external terminals excited by a steady-state sinusoidal source with $s = j\omega$ and $V_2 = 0$, Eq. (3.13) becomes

$$V_1 I_1^* = 2P_d + j2\omega(W_m - W_e) \qquad (3.14)$$

with

$$P_d = \frac{1}{2}\sum_{j,k=1}^{n} R_{jk} I_j I_k^* \quad = \text{average power dissipated in resistors of network}$$

$$W_m = \frac{1}{2}\sum_{j,k=1}^{n} L_{jk} I_j I_k^* \quad = \text{total average magnetic energy stored in inductors}$$

$$W_e = \frac{1}{2\omega^2}\sum_{j,k=1}^{n} \frac{1}{C_{jk}} I_j I_k^* = \text{total average electrical energy stored in capacitors}$$

These functions are said[4] to be positive real since they can never become negative. Equation (3.14) expresses the power balance between the input on one hand and the real dissipated power and reactive stored 'power' in the network on the other. In the following chapter, it is seen to lead to fundamental definitions of the filter properties of selectivity and sensitivity.

Reverting to s notation and dividing Eq. (3.13) throughout by $I_1 I_1^* = |I_1|^2$, the driving-point impedance of the one-port with $V_2 = 0$ is related to the energy functions by

$$Z(s) = \frac{V_1}{I_1} = \frac{1}{|I_1|^2}\left(P_{d0} + sW_{m0} + \frac{W_{e0}}{s}\right) \qquad (3.15)$$

Of course, this is not an explicit expression in s since the energy functions themselves are dependent on s via I_j. However, P_{d0}, W_{m0}, and W_{e0} are always real and non-negative while $|I_1|^2$ must be positive. Hence, two basic properties emerge which are said to define *positive-real functions* (PRF's):

(a) $Z(s)$ is real for s real
(b) the real part $\operatorname{Re} Z(s) \geqslant 0$ for $s \geqslant 0$

The driving-point impedance of a passive bilateral network is thus seen always to be positive real and the same is true for admittance by dividing Eq. (3.13) throughout by $V_1 V_1^*$.

As the second property is extremely difficult to test, equivalent sets of necessary and sufficient conditions are available for routine application. These are now listed with reference to an immittance function having the general bi-polynomial form of Eq. (2.1), their derivation and substantiation being available[1] elsewhere:

1. All polynomial coefficients are real and positive (a consequence of property (a)).
2. The scale factor a_0/b_0 is real and positive.
3. The highest degrees of the numerator and denominator polynomials differ at most by unity.
4. The lowest degrees of the numerator and denominator polynomials differ at most by unity.
5. There are no missing terms in the numerator and denominator polynomial unless all even or all odd terms are missing.
6. Poles and zeros are real or occur in conjugate pairs.
7. Poles and zeros on the imaginary axis are simple with real positive residues.
8. There are no poles in the right half-plane, i.e., the function is analytic which is consequent on the denominator polynomial being Hurwitz.[1]
9. The real part of the function is non-negative along the entire $j\omega$ axis, i.e., $\operatorname{Re} Z(j\omega) \geqslant 0$ for all ω.

This is the normal order of testing since 1 to 5 require mere inspection but 6 and 7 require factorization, 8 requires a continued-fraction development, and 9 requires either graphical or Sturm-function evaluation.

3.3.2 Transfer functions

Turning now to the two-port case, replace the voltage matrix of Eq. (3.12) by its equivalent in terms of impedance parameters, so that

$$[I^*]^t[z][I] = z_{11}I_1I_1^* + 2z_{12}\mathrm{Re}(I_1I_2^*) + z_{22}I_2I_2^* \quad (3.16)$$

$$= P_{d0} + sW_{m0} + \frac{W_{e0}}{s}$$

with the notation of Fig. 3.4 and parameter description by equation couplet

$$V_1 = z_{11}I_1 + z_{12}I_2 \quad \text{and} \quad V_2 = z_{21}I_1 + z_{22}I_2 \quad (3.17)$$

Since Eq. (3.16) must be a positive-real function and the matrix of a positive-definite quadratic form is also positive definite, then the important conclusion emerging is that the impedance matrix of a passive bilateral two-port is positive real. The same is true for the admittance matrix.

A pole of any of the impedance parameters is seen also to be a pole of the overall energy function in Eq. (3.16). Since this is positive real, then its residue k at a pole on the $j\omega$ axis must be real and non-negative, i.e.,

$$k = k_{11}x_1^2 + 2k_{12}x_1x_2 + k_{22}x^2 \geqslant 0,$$

where k_{11} is the corresponding residue of z_{11}, etc. This gives rise to the so-called *residue conditions*:

$$k_{11} \geqslant 0, \quad k_{22} \geqslant 0 \quad \text{and} \quad k_{11}k_{22} - k_{12}^2 \geqslant 0 \quad (3.18)$$

Similar conditions hold for the real parts of the impedance parameters. The importance of Eq. (3.18) in network synthesis can be judged by the fact that when k_{12} is non-zero, then both k_{11} and k_{22} must also be non-zero. This means that if z_{12} has a pole on the $j\omega$ axis, then so must z_{11} and z_{22}. However, each of the driving-point functions may have an exclusive or *private pole* realizable by series-arm impedances, as in Fig. 3.5(a), for which

$$z_{11} = z'_{11} + z_1, \quad z_{22} = z'_{22} + z_2, \quad z_{12} = z'_{12}$$

For admittance functions, additional shunt arms result as in Fig. 3.5(b).

Fig. 3.5. (a) Two-port network with series-arm elements; (b) two-port network with shunt-arm elements

The realizability of a transfer function $T(s)$ necessitates expression in terms of immittance parameters. For example, Eq. (3.17) gives

$$T(s) = \left.\frac{V_2}{V_1}\right|_{I_2=0} = \frac{z_{12}}{z_{11}} \qquad (3.19)$$

while analogously, or by the matrix-conversion table of Appendix 5,

$$T(s) = \frac{-y_{12}}{y_{22}} \qquad (3.20)$$

By such considerations, a number of transfer-function properties can be listed for the general LLFPBN:

1. There are no poles of z_{12}, y_{12}, or T in the right half-plane.
2. There are no poles of T at zero or infinity.
3. Poles of T on the $j\omega$ axis are simple with imaginary residues.
4. Poles of z_{12} and y_{12} on the $j\omega$ axis are simple with real residues.
5. The immittance parameters satisfy the residue condition at all poles on the $j\omega$ axis.
6. Zeros may be non-simple and occur anywhere in the s plane.
7. Immittance parameters satisfy the real-part conditions.
8. The degree of the numerator of T is not greater than the degree of the denominator.

When the specific structure of the network is taken into account, the properties need to be reconsidered. Two broad categories can be distinguished: the *unbalanced network* with one terminal common to both input and output

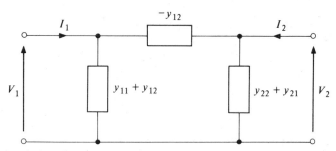

Fig. 3.6. π equivalent of two-port network

and the *balanced network* with no common terminal. In the case of the former, T–π conversions enable the pi equivalent to be drawn as in Fig. 3.6. Notice that all coefficients of admittance branches $-y_{12}$, $y_{11} + y_{12}$, $y_{22} + y_{21}$ are positive. Expressing the parameters as

$$-y_{12} = \frac{\sum_{i=0}^{n} a_i s^i}{Q(s)}, \quad y_{11} = \frac{\sum_{i=0}^{n} b_i s^i}{Q(s)}, \quad y_{22} = \frac{\sum_{i=0}^{n} c_i s^i}{Q(s)} \qquad (3.21)$$

where $Q(s)$ is a divisor polynomial, then the branch admittances become

$$-y_{12} = \frac{\sum_{i=0}^{n} a_i s^i}{Q(s)}, \quad y_{11} + y_{12} = \frac{\sum_{i=0}^{n} (b_i - a_i)s^i}{Q(s)}, \quad y_{22} + y_{12} = \frac{\sum_{i=0}^{n} (c_i - a_i)s^i}{Q(s)} \quad (3.22)$$

from which the *Fialkow–Gerst* conditions are stated[1,4] as

$$a_i \geqslant 0, \quad b_i \geqslant a_i, \quad c_i \geqslant a_i \quad (3.23)$$

being valid if no common factors have been cancelled in the numerator and denominator of the branch admittance functions.

3.4 Synthesis procedures

In synthesis, it is required to determine a network structure which realizes a given filter performance. When the latter is presented as a frequency response, the approximation techniques of Chapter 2 are first applied. The properties of the resulting function have been discussed in Sec. 3.1 and the conditions for realizability in Sec. 3.3. Methods of formal synthesis are now outlined with respect to RC realizations only. Similar developments are available for LC and RL networks or, alternatively, these may be obtained by the transformations stated in Sec. 3.2. Again, the cases of immittance and transfer functions are considered separately. An important feature of synthesis as distinct to analysis is that a resulting network realization for a given function is not unique, and alternative realizations exist which may have more desirable properties. The actual values of components obtained require subsequent denormalization with respect to operating frequency and impedance level (see Appendix 1).

3.4.1 Synthesis of driving-point immittance functions

The synthesis of a one-port positive-real driving-point function devolves to its decomposition into simpler functions which are recognizable as network immittances. These are then 'removed', leaving a new lower-order function which may be further reduced.

Partial-fraction synthesis The factorized RC-impedance function

$$Z(s) = \frac{K(s + a_1)(s + a_2)\cdots}{s(s + \sigma_1)(s + \sigma_2)\cdots} \quad (3.24)$$

can be expressed as a partial-fraction expansion

$$Z(s) = \frac{k_0}{s} + \frac{k_1}{s + \sigma_1} + \frac{k_2}{s + \sigma_2} + \cdots + k_\infty \quad (3.25)$$

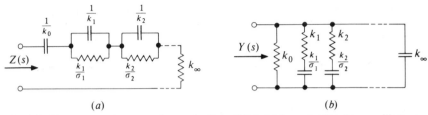

Fig. 3.7. (*a*) Foster series-connection realization of *RC* driving-point immittance; (*b*) Foster parallel-connection realization of *RC* driving-point immittance

where k_0, k_1, k_2 are positive-real residues of corresponding real poles 0, σ_1, σ_2 and k_∞ is a constant at infinity. Each term in Eq. (3.25) can be recognized as the impedance of the simple parallel-*RC* circuit type shown in Table 3.1 and the overall function realized by the *Foster series connection* of Fig. 3.7(*a*). The single capacitor represents a pole at the origin and the single resistor is absent if there is a zero at infinity. Derivation of the *Foster parallel connection* of Fig. 3.7(*b*) requires expansion of $Y(s)/s$ and recognition of terms as simple series *RC* circuits from Table 3.1.

Continued-fraction synthesis Realization of Eq. (3.24) can also be carried out by proceeding alternately on an impedance and then an admittance basis so as always to remove a constant or a pole at infinity. This necessitates inversion of the remainder function at each step so that the impedance function is developed as a finite continued fraction

$$Z(s) = R_1 + \cfrac{1}{sC_1 + \cfrac{1}{R_2 + \cfrac{1}{sC_2 + \cdots}}} \tag{3.26}$$

The resulting *first Cauer ladder structure* is presented in Fig. 3.8(*a*), the nature of the impedance function at infinity determining the first element and at zero the last element. The *second Cauer ladder* of Fig. 3.8(*b*) emerges by successive removal of poles at the origin. In each case, the number of inversions is obviously equal to the order of the rational function and, consequently, Cauer

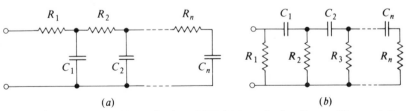

Fig. 3.8. (*a*) First Cauer ladder realization of *RC* driving-point immittance; (*b*) second Cauer ladder realization of *RC* driving-point immittance

ladders like the Foster connections are canonical realizations, i.e., have a minimum number of elements.

3.4.2 Synthesis of transfer functions

Attention is here confined to *exact synthesis* techniques as distinct to the simpler but approximate *image-parameter* methods of two-port synthesis originally based on analogy with the transmission line. The voltage transfer ratio of the network in Fig. 3.4 is expressed in terms of immittance parameters by Eqs. (3.19) and (3.20), affording considerable flexibility in realization since three parameters are necessary to completely specify a two-port section. In this case, it is the zeros of transmission rather than the poles which influence behaviour. Hence the procedures for realizing transfer functions reduce to the problem of synthesizing the associated driving-point immittance function in such a way that the zeros of transmission are also realized. The manner in which these zeros are produced depends on whether the transfer function is *minimum-phase*, i.e., it has no zeros in the right half of the complex plane. The latter condition is sufficient but not necessary to permit realization as an *unbalanced network* structure. Two methods particularly suitable for RC-active filter design are now considered.

RC ladder If the transmission zeros are restricted to the negative-real axis, then a simple RC ladder suffices. The zeros are then produced by breaking the transmission path with an open circuit in a series arm or a short circuit in a shunt arm (see Fig. 3.9), these being effected by RC-resonant and antiresonant combinations of elements of the type in Table 3.1. The realization procedure[5] is an extension of that described in Sec. 3.4.1 once the driving-point and transfer immittances have been established from the given transfer function. The driving-point parameter is developed incorporating the appropriate transmission zeros in the process by successive zero shifting and removal. The network structure resulting is not unique since, having assigned the zeros of $T(s)$ as zeros of z_{21} or $-y_{21}$ and the poles of $T(s)$ as the zeros of z_{11} and y_{22}, there is freedom in the choice of poles for the immittance functions (provided that they alternate with the zeros and satisfy criteria at the lowest and highest frequencies). The same technique is applicable to the LC ladder by the transformation of Sec. 3.2.1, the zeros now being restricted to the $j\omega$ axis. Here, however, it is obviously not possible to incorporate resistive terminations as part of the two-port network.

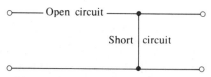

Fig. 3.9. Ladder element with zero-producing branches

Parallel-RC ladder Complex zeros may be realized by resorting to a multiple-path configuration. The admittance parameters pertaining to an open-circuit voltage transfer function are decomposed so that the transmission zeros of each constituent lie on the negative-real axis, then each component is realized as an *RC* ladder. The ladders are then paralleled as in Fig. 3.10 to

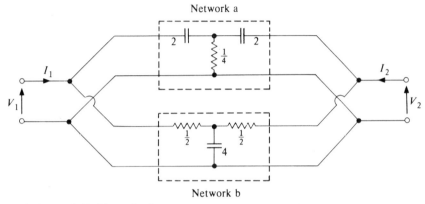

Fig. 3.10. Parallel-ladder realization

complete the realization. The actual procedure is to develop the admittance parameters for the voltage transfer of Eq. (2.1) as

$$y_{22} = \frac{D(s)}{Q(s)} \quad \text{and} \quad -y_{21} = \frac{N(s)}{Q(s)}$$

$$= \frac{a_0 + a_1 s + a_2 s^2 + \cdots + a_n s^n}{Q(s)}$$

$$= \frac{a_0}{Q(s)} + \frac{a_1 s}{Q(s)} + \frac{a_2 s^2}{Q(s)} + \cdots + \frac{a_n s^n}{Q(s)} \quad (3.27)$$

in which the coefficients $a_0, a_1, \cdots, a_n$ are positive and the transmission zeros are all at 0 or ∞. This indicates that the number of ladders required is one more than the order of the function. Less extravagant realizations are possible by choosing a different decomposition. It is necessary to adjust the constant multiplier achieved for each y_{12} so as to restore the original overall function. This is achieved[4] by scaling the admittance level of, for example, the ith ladder by

$$\left(\frac{a_i}{B_i}\right) \Big/ \left(\sum_{i=0}^{n} \frac{a_i}{B_i}\right)$$

where the realized parameter $y'_{12i} = B_i \cdot y_{12i}$. All capacitors are scaled by this factor and all resistors by its inverse before interconnecting the network.

PASSIVE NETWORKS

Example 3.1 Synthesize the voltage transfer function $(s^2 + 1)/(s^2 + 4s + 1)$ as a parallel-RC ladder structure.

SOLUTION Selecting divisor polynomial $Q(s) = s + \sigma$ gives short-circuit admittance parameters

$$-y_{21_a} = \frac{s^2}{s + \sigma}, \quad -y_{21_b} = \frac{1}{s + \sigma}, \quad \text{and} \quad y_{22} = \frac{s^2 + 4s + 1}{s + \sigma}$$

The transfer admittances are readily recognizable as belonging to simple T sections listed in Table 3.2. The corresponding driving-point admittance is

$$y_{22_a} + y_{22_b} = \frac{2s(s/2 + \sigma)}{s + \sigma} + \frac{2s/\sigma + 1}{s + \sigma} = \frac{s^2 + 2(\sigma + 1/\sigma)s + 1}{s + \sigma}$$

dictating $\sigma = 1$. The realization is given in Fig. 3.10 being, in fact, the well-known null-adjusted symmetrical twin-T network. This simple case also clarifies the synthesis procedure, though it can be readily generalized for arbitrary coefficients.

Shortcomings of the parallel-ladder approach are the often excessive number of elements, the inability to realize transfer impedances, and the low transmission gain. While the latter is of less importance in active configurations and can be overcome by use of the *Fialkow–Gerst tree structure*,[4] there is the additional disadvantage of no single-element control of the position of each zero. The impossibility of independent adjustment can render the tuning problem (see Chapter 9) more difficult. The *Dasher cascade synthesis* procedure allows control of complex-zero-producing sections in a ladder fashion but at the expense of more computation. There is as yet no solution to the general synthesis problem for three prescribed RC-immittance parameters. Positive-real zeros can be realized in an all-passive RC structure only by resort to a balanced form which does not allow a common earth line.

3.4.3 Coefficient-matching technique

The formal synthesis methods outlined in the preceding section usually give the circuit designer no hint of the final topology. The latter is not unique, but alteration of the resulting realization is invariably not a simple matter. In most active-filter situations, an alternative pre-analysis approach of coefficient matching[5,6] is more convenient. This assumes a given network structure at the outset for which the transfer function is determined by conventional analysis[2] with coefficients dependent on the element values. By equating like coefficients with the function to be realized, synthesis devolves to the problem of solving a

Table 3.2 Short-circuit transfer admittances of common RC two-port networks

Type of function	$-y_{21}$	Circuit	Element values
Lowpass	$\dfrac{a}{s+\sigma}$		$G = 2a/\sigma$ $C = 4a/\sigma^2$
Lowpass	$\dfrac{a}{(s+\sigma_1)(s+\sigma_2)}$		$G_1 = 2a/\sigma_2(\sigma_1 - \sigma_2)$ $G_2 = a/\sigma_2^2$ $C = 2a/\sigma_2^2(\sigma_1 - \sigma_2)$
Highpass	$\dfrac{as^2}{s+\sigma}$		$G = 4a\sigma$ $C = 2a$
Highpass	$\dfrac{as^2}{(s+\sigma_1)(s+\sigma_2)}$		$G_1 = a$ $G_2 = \dfrac{(\sigma_1-\sigma_2)^2 \pm (\sigma_1+\sigma_2)\sqrt{(\sigma_1+\sigma_2)^2 - 8\sigma_1\sigma_2}}{(2\sigma_1\sigma_2/a)}$ $C = \dfrac{(\sigma_1+\sigma_2) \pm \sqrt{(\sigma_1+\sigma_2)^2 - 8\sigma_1\sigma_2}}{(2\sigma_1\sigma_2/a)}$
Bandpass	$\dfrac{as}{s+\sigma}$		$G = a$ $C = a/\sigma$

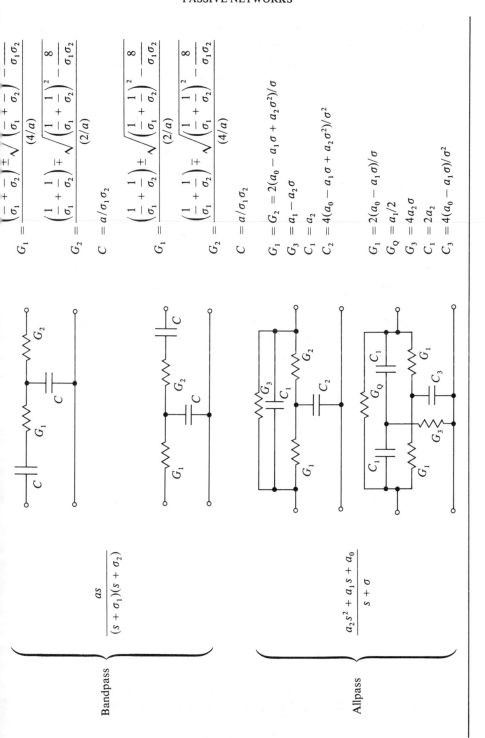

set of simultaneous algebraic equations for the normalized values of components. This procedure is illustrated by several examples in later chapters.

As a technique, it is particularly effective for simple second-order transfer functions which, as is to be seen in Chapters 4 and 9, form the basis for much of active-filter design. The initial choice of topology has practical advantages and is carried out on the basis of experience or by reference to tabulated results as in Table 3.2, which presents short-circuit transfer admittances and element values for commonly occurring passive RC two-ports. Furthermore, it almost always allows degrees of freedom in design on account of the non-unique solutions to the simultaneous equations. However, the non-linear nature of these equations renders solution difficult for high orders and severely limits the technique when there are more than three finite poles and zeros.

Ladder cumulants The ladder structures emerging from the synthesis procedures described in Secs. 3.4.1 and 3.4.2 are particularly useful practical forms. It is convenient to characterize the general ladder by a network-function determinant with straightforward expansion and recursion rules which enable development of any special case prior to coefficient-matching design. Alternate mesh and nodal analysis of the general ladder with no node bridging, as in Fig. 3.11, leads to the set of equations

$$\begin{aligned}
V_0 &= z_1 i_1 + V_1 \\
0 &= -i_1 + Y_1 V_1 + i_2 \\
0 &= \quad\quad -V_1 + z_2 i_2 + V_2 \\
&\;\vdots \\
0 &= \quad\quad\quad\quad\quad\quad \cdots -i_n + Y_n V_n
\end{aligned} \quad (3.28)$$

The determinant of the matrix on the right-hand side

$$\Delta = \begin{vmatrix}
z_1 & 1 & 0 & 0 & \cdots & 0 & 0 & 0 \\
-1 & Y_1 & 1 & 0 & \cdots & 0 & 0 & 0 \\
0 & -1 & z_2 & 1 & \cdots & 0 & 0 & 0 \\
0 & 0 & -1 & Y_2 & \cdots & 0 & 0 & 0 \\
\vdots & \vdots & \vdots & \vdots & & \vdots & \vdots & \vdots \\
0 & 0 & 0 & 0 & \cdots & -1 & z_n & 1 \\
0 & 0 & 0 & 0 & \cdots & 0 & -1 & Y_n
\end{vmatrix} \quad (3.29)$$

can be entirely represented by the elements of the principal diagonal and is termed the network *cumulant*[8] or simple *continuant*[5] denoted by

$$K(z_1, Y_1, z_2, Y_2, \cdots, z_n, Y_n) = \Delta \quad (3.30)$$

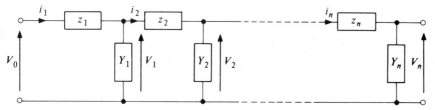

Fig. 3.11. General *n*-stage ladder network

Applying Cramer's rule,

$$V_n = \frac{\Delta_{nn}}{\Delta} \cdot V_0 \qquad (3.31)$$

where Δ_{nn} is the determinant obtained by replacing the last column of Eq. (3.29) by the left-hand side of Eq. (3.28) and extracting V_0. Since this determinant can be developed as

$$\Delta_{nn} = (-1)^{n-1} V_0$$

then the voltage gain of the ladder is given by

$$\frac{V_n}{V_0} = \frac{1}{\Delta} = \frac{1}{K(z_1, Y_1, z_2, Y_2, \cdots, z_n, Y_n)} \qquad (3.32)$$

Rules for the expansion of the cumulant are available,[8] obviating the need for evaluation of determinants in each specific case. The open-circuit voltage transfer ratio for the widely used three-section ladder is presented in Table 3.3 in terms of general admittances. The transfer ratios for the general twin-T (introduced in Ex. 3.1 as a parallel ladder) and bridged two-section ladder are also listed. These forms cater for almost every combination of passive elements used in second-order active filters by appropriate substitution of admittance terms and simplification where necessary.

The RC twin-T network When the structure of Fig. 3.10 is generalized by means of Table 3.3, the resulting twin-T of Fig. 3.12 is basically[9] a third-order

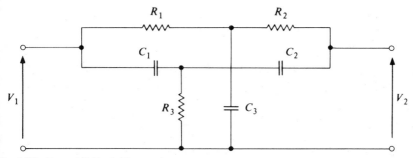

Fig. 3.12. General *RC* twin-T network

Table 3.3 Open-circuit voltage transfer ratios for general unbalanced passive network configurations

Network form	Open-circuit voltage transfer ratio V_2/V_1
Three section ladder	$$\frac{y_1 y_2 y_3}{y_2(y_1+Y_1)(y_3+Y_3)+(y_1+y_2+Y_1)(y_3 Y_2+y_3 Y_3+Y_2 Y_3)}$$
Twin-T	$$\frac{Y_1 Y_2 (Y_3+Y_4+Y_6)+Y_3 Y_4 (Y_1+Y_2+Y_5)}{Y_2(Y_1+Y_5)(Y_3+Y_4+Y_6)+Y_4(Y_3+Y_6)(Y_1+Y_2+Y_5)}$$
Bridged ladder	$$\frac{Y_1 Y_2 + Y_6 (Y_1+Y_2+Y_5)}{Y_2(Y_1+Y_5)+(Y_4+Y_6)(Y_1+Y_2+Y_5)}$$

network with voltage transfer function

$$\frac{V_2}{V_1} = \frac{R_1R_2R_3C_1C_2C_3s^3 + R_3(R_1+R_2)C_1C_2s^2 + R_3(C_1+C_2)s + 1}{\begin{array}{c}R_1R_2R_3C_1C_2C_3s^3 + [R_1R_3C_3(C_1+C_2) + R_3(R_1+R_2)C_1C_2 \\ + R_1R_2C_2C_3]s^2 + [R_1(C_2+C_3) + R_3(C_1+C_2) + R_2C_2]s + 1\end{array}}$$

(3.33)

Under the null-adjusted condition $R_1R_2C_3 = R_3(R_1+R_2)(C_1+C_2)$, factorizing and cancelling the first-order term $[s + 1/R_3(C_1+C_2)]$ gives the second-order notch function of the elliptic type of Eq. (2.16) as

$$\frac{V_2}{V_1} = \frac{(s/\omega_0)^2 + 1}{(s/\omega_0)^2 + (1/Q)(s/\omega_0) + 1} \qquad (3.34)$$

where

$$\omega_0 = \sqrt{\frac{C_1+C_2}{R_1R_2C_1C_2C_3}} \quad \text{and} \quad Q = \frac{1}{\omega_0} \cdot \frac{1}{R_1C_3 + (R_1+R_2)C_2} \qquad (3.35)$$

The nature of Eq. (3.34) makes this network of prime importance in the design of many active filters to be discussed in later chapters. In order to confine component spread to dependence on a single factor k, it is usual to make the identities $R_2 = kR_1$, $R_3 = (k/(k+1))R_1$, $C_2 = C_1/k$, and $C_3 = (1 + 1/k)C_1$ constituting the potentially symmetric[9] twin-T network. The uniform structure of Ex. 3.1 emerges for $k = 1$. The remarkable properties of the twin-T are somewhat offset, however, by difficulties such as tuning (see Chapter 9), as may be seen by the interdependencies of Eqs. (3.35).

3.5 Distributed-parameter networks

The development of integrated-circuit technology has allowed the fabrication of networks having distributed as well as lumped parameters. In practice, even discrete elements themselves display distributed parasitic effects. Consequently, the theory of electrical transmission lines is of continuing relevance in the analysis of microelectronic circuits. Thin-film deposition techniques (see Sec. 9.1) evolving from printed strip lines in the 1950's introduced a new class of circuit component illustrated in Fig. 3.13. Here, the resistive and dielectric layers are formed such that their electromagnetic fields interact, thereby permitting circuit functions not realizable using a finite number of discrete resistors and capacitors. The films are deposited on an inert substrate and conductive connections made as indicated. The sheet thicknesses are enlarged for pictorial clarity while fabrication processes are discussed in the perspective of other technologies in Sec. 9.3.

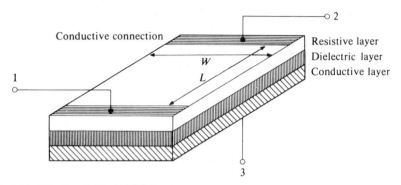

Fig. 3.13. Thin-film distributed-RC structure

In order to derive network parameters for the distributed-RC or $\overline{RC}$ structure, a one-dimensional current flow is assumed. For a three-terminal uniform structure with lossless dielectric and negligible series inductance, the schematic circuit symbol is shown in Fig. 3.14. Treating each infinitesimal section of the line as a single network element with elemental series resistance

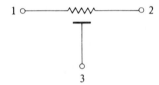

Fig. 3.14. Circuit symbol for uniform distributed-RC section

$r_0 \cdot \delta x$ and shunt capacitance $c_0 \cdot \delta x$, where x is the distance variable, then the voltage difference equation for Fig. 3.15 is

$$V - r_0 \cdot \delta x \cdot i = V + \delta V \tag{3.36a}$$

and the current difference equation

$$i - sc_0 \cdot \delta x (V + \delta V) = i + \delta i \tag{3.36b}$$

where r_0 is the resistance per unit length, c_0 is the capacitance per unit length, V is the voltage along the line and i is the current along the line. Neglecting

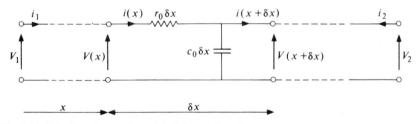

Fig. 3.15. Distributed-network model of elemental sections

PASSIVE NETWORKS

difference products gives, in the limit, the so-called 'Telegraphers' equations':

$$\frac{dV}{dx} = -r_0 \cdot i \qquad (3.37a)$$

and

$$\frac{di}{dx} = -sc_0 \cdot V \qquad (3.37b)$$

Combining these gives the second-order voltage differential equation

$$\frac{d^2V}{dx^2} = sr_0 c_0 V \qquad (3.38)$$

which, being in the normal form, has solution

$$V = ae^{-\gamma x} + be^{+\gamma x} \qquad (3.39)$$

since $\gamma^2 = sr_0 c_0$ is independent of x. Evaluating constants a and b by applying boundary conditions at the input ($x = 0$) and output ($x = L$, the length of the finite line), then substituting in the current solution obtained from Eqs. (3.37a) and (3.37b), leads to

$$i = -\frac{1}{r_0} \cdot \frac{dV}{dx} = \frac{\gamma}{r_0} \left[V_1 \frac{\cosh \gamma(L-x)}{\sinh \gamma L} - V_2 \frac{\cosh \gamma x}{\sinh \gamma L} \right] \qquad (3.40)$$

From this, the conventional admittance-parameter definitions applied at the terminals give the two-port matrix

$$[y] = \frac{\gamma}{r} \begin{bmatrix} \coth \gamma L & -\operatorname{cosech} \gamma L \\ -\operatorname{cosech} \gamma L & \coth \gamma L \end{bmatrix} \qquad (3.41)$$

The corresponding impedance parameters can be derived by developing Eq. (3.39) or by the matrix conversions of Appendix 5. The same hyperbolic functional forms become evident[10] when the distributed line is approximated by a cascade of lumped sections as in the ladder structure of Fig. 3.11.

Extension of the above analysis allows characterization of multilayer and non-uniform distributed sections[11] with attendant advantages in performance together with further economy in substate area. The inherently low component count of distributed realizations entails fewer interconnections and consequently affords improved simplicity and reliability. However, these practical advantages are offset by the complexity associated with the irrational nature of the parameters in Eq. (3.41). These render analysis difficult and synthesis impossible except by transformation of the frequency variable. Furthermore, the triple-layer thin-film morphology of Fig. 3.13 cannot be readily trimmed by the methods to be discussed in Chapter 9. It is in combination with lumped elements that distributed $\overline{RC}$ networks offer most potential by a compromise between component count and tuning facility. Perhaps the most useful of such arrangements is that of Fig. 3.16 in which, at a particular frequency, the voltage V_R across leg resistor R_1 is equal in amplitude and opposite in phase to the voltage V_L at the output end of the line. By addition of voltage vectors, the signal V_2 at the second port then becomes zero. That is, there is a null[12] at this frequency and hence the circuit constitutes a notch filter. As such, it may be usefully employed as an analogous replacement in all active network configurations incorporating the RC twin-T section.

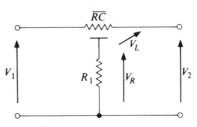

Fig. 3.16. Distributed-lumped notch circuit

3.6 Limitations of passive networks

Passive LCR filters can have natural frequencies anywhere in the left half of the complex-frequency plane and as such are widely used, especially at high frequencies, whereas in and below the audio band the demands on components become severe. In particular, the physical size and weight of the inductors is prohibitive in addition to their large dissipation factors complicating synthesis (see Appendix 2), their unsuitability for integrated-circuit realization, and their non-linear core-saturation characteristics, to say nothing of their expense. On the other hand, passive RC filters have been seen to have natural frequencies restricted to the negative-real axis. This limits the quality of approximation to an ideal response.

In order to dispense with inductors, yet retain the range of transfer-function capability, it is necessary to introduce active elements which (fortuitously) are ideally suited to microelectronic fabrication (see Chapter 9). This effectively allows negative components, and hence the possibility of natural frequencies anywhere in the complex plane. So long as stability is assured, it might appear that there is no limit to the sharpness of response achievable by this means. Factors limiting performance are discussed in the following chapter.

References

1. M. E. Van Valkenburg, *An Introduction to Modern Network Synthesis*, Wiley, New York, 1960.

2. M. E. Van Valkenburg, *Network Analysis*, 3rd ed., Prentice Hall, Englewood Cliffs, N.J., 1974.
3. L. Weinberg, *Network Analysis and Synthesis*, McGraw-Hill, New York, 1962.
4. N. Balabanian, *Network Synthesis*, Prentice Hall, Englewood Cliffs, N.J., 1958.
5. E. S. Kuh and D. O. Pederson, *Principles of Circuit Synthesis*, McGraw-Hill, New York, 1959.
6. S. K. Mitra, *Analysis and Synthesis of Linear Active Networks*, Wiley, New York, 1969, Chap. 3.
7. E. A. Guillemin, *Introductory Circuit Theory*, Wiley, New York, 1953.
8. J. L. Herrero and G. Willoner, *Synthesis of Filters*, Prentice Hall, Englewood Cliffs, N.J., 1966.
9. G. S. Moschytz, 'A general approach to twin-T design and its application to hybrid integrated linear active networks', *Bell Systems Tech. J.*, **49** (7), 1105–1149, 1970.
10. M. S. Ghausi and J. J. Kelly, *Introduction to Distributed-parameter Networks*, Holt, Rinehart, and Winston, New York, 1968.
11. P. Bowron, 'Matrices for dual representation of four-terminal non-uniform transmission lines', *Radio Electron. Engr*, **43** (9), 537–544, 1973.
12. H. C. Lin, *Integrated Electronics*, Holden-Day, San Francisco, 1967.

4. Selectivity, sensitivity, and stability

The limitations of passive filter realizations have become apparent in the preceding chapter: inductors are physically inconvenient while RC structures with natural frequencies restricted to the negative-real axis of the complex-frequency plane are severely restricted in the quality of approximation to an ideal response. This shortcoming can be circumvented by the introduction of active elements to be described in Chapter 5. The resultant RC-active networks are capable of realizing natural frequencies anywhere in the complex plane.

This new freedom, however, is attended by the appearance of further problems: namely, sensitivity and stability. The techniques invoked to achieve response selectivity are inherently associated with these two problems. Whereas stability is assured for the all-passive realizations of Chapter 3, this is not so for active networks which may have natural frequencies straying into the right half-plane on account of changes in sensitive parameters. Even if these do not lead to instability, they may produce large variations in the filter response, possibly resulting in the filter failing to meet its specification in the face of temperature drift, ageing, or replacement of components. Hence, high sensitivity implies that component tolerances must be tightly controlled otherwise deviations from the nominal values will cause intolerable deviations in the filter response. Furthermore, different philosophies of activating RC sections lead to different sensitivity performances. In order to compare these, appropriate criteria are developed first for a single critical parameter and then for the more realistic case of multiple parameters, thereby permitting the effects of manufacturing tolerances and environmental factors to be assessed.

4.1 Selectivity

Selectivity is a familiar term in telecommunications engineering and may be defined[1] as 'the ability of a system to distinguish between desired and undesired signals.' In the frequency domain, it is a parameter fundamental to the problem of filtering, yet the diversity of filter functions implies that no single comprehensive and convenient criterion can be formulated. For band-emphasis frequency responses, it is obviously appropriate to measure performance in terms of selectivity. However, in the context of band elimination, it is perhaps more meaningful[2] to speak of 'rejectivity'. Highpass and lowpass frequency responses are, of necessity, usually[3] described in terms of sharpness of cut-off, i.e., the rate at which the amplitude changes in the transition-frequency zone (see Chapter 2). Hence, a variety of definitions is necessary for comparison of non-complex filter structures, as described in the following subsections.

4.1.1 The meaning of Q

The concept of quality factor or Q is one of the most equivocal in engineering, there being[1] no general agreement on what it is intended to represent. Even its

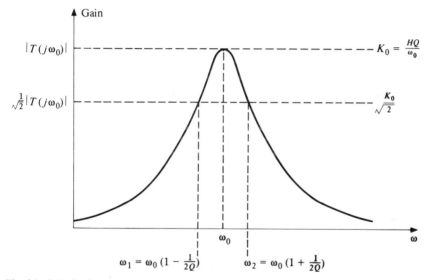

Fig. 4.1. Selective frequency response

historical origins are vague and, while the notion of selectivity has been familiar since the inception of radio communication, its definition in circuit and filter theory has been haphazard. It is customary to measure the sharpness of a resonant response by taking the reciprocal of the half-power bandwidth. The so-called *'classical' selectivity* definition is illustrated in Fig. 4.1, for which

$$Q = \frac{\omega_0}{\omega_2 - \omega_1} = \frac{1}{\Omega_2 - \Omega_1} \qquad (4.1)$$

where ω_1 and ω_2 are the lower and upper angular frequencies, respectively, at which the modulus of transfer function $T(j\omega)$ is at half power, while ω_0 is the centre or resonant angular frequency.

By considering the reactance slope of the steady-state form of Eq. (3.15) with the notation of Eq. (3.14), the definition of Eq. (4.1) at the input port of a passive network can be shown[4] to approach

$$Q \to \frac{\omega |\partial/\partial\omega[2\omega(W_m - W_e)]|}{2P_d}\bigg|_{\omega=\omega_0} \qquad (4.2)$$

For the simple series or parallel RLC circuit tuned to narrow relative bandwidth in the vicinity of the resonant frequency, Eq. (4.2) can readily be seen to reduce to

$$Q \simeq \frac{\omega(W_m + W_e)}{P_d}\bigg|_{\omega=\omega_0} \qquad (4.3)$$

From this may be recognized the energy formulation of selectivity, so widely applied in physical systems:

$$Q_e = 2\pi \frac{\text{maximum energy stored}}{\text{energy dissipated per cycle}} \qquad (4.4)$$

However, far from furnishing a universal definition, the last equation merely adds to the ambiguity surrounding Q. The restrictions involved should be constantly remembered, and any attempt to apply energy principles to general circuits containing more than one inductor or capacitor should strictly resort to the form of the frequency-derivative of the difference in stored energies as presented in Eq. (4.2). Of course, in RC-active networks, one is concerned only with capacitive reactances and their interchange of energy with active elements.

Q is also useful as a measure of the quality or storage of a reactive element. For example, an inductive coil is not an ideal reactance on account of the losses associated with winding resistance and core permeability. However, in network analysis, it is convenient to maintain the concept of pure inductance and represent all dissipative losses by means of an equivalent resistance either in series (R_s in Fig. 4.2(a)) or in parallel (R_p in Fig. 4.2(b)) with the effective

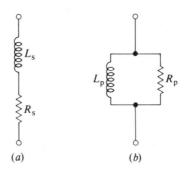

Fig. 4.2. Circuit models for non-ideal inductors. (a) series equivalent; (b) parallel equivalent

reactance of the actual element. The energy definition of Eq. (4.3) is applicable here and for the series representation gives the quality factor as

$$Q_s = \omega L_s / R_s \qquad (4.5a)$$

while for the transformed parallel equivalent

$$Q_p = R_p / \omega L_p \qquad (4.5b)$$

where

$$R_p = R_s(1 + Q_s^2) \quad \text{and} \quad L_p = L_s(1 + 1/Q_s^2).$$

In the audio band, the quality factor of a wound coil is usually limited to less than a thousand and considerably lower at higher frequencies. Losses in a capacitor are normally smaller: its Q factor is given by the ratio of susceptance to conductance, expressed as the reciprocal of the loss tangent (tan δ).

The above energy definitions are usually awkward to evaluate for general circuits, but, for systems described by second-order functions, a particularly simple figure of merit for selectivity is apparent. First hinted in differential-equation terms, it was formulated by Morris[5] with suitable scaling of the denominator coefficients of the transfer or transmission function. For the left-half-plane pole pair $-\sigma_i \pm j\omega_i$ in Fig. 4.3, this becomes

$$\begin{aligned} T(s) &= \frac{N(s)}{b_2 s^2 + b_1 s + b_0} \\ &= \frac{N(s)}{b_2(s + \sigma_i - j\omega_i)(s + \sigma_i + j\omega_i)} = \frac{1}{b_2} \frac{N(s)}{s^2 + 2\sigma_i s + (\sigma_i^2 + \omega_i^2)} \\ &= \frac{1}{b_0} \frac{N(s)}{(s/\omega_0)^2 + (1/Q_m)(s/\omega_0) + 1} \end{aligned} \qquad (4.6)$$

where, for high selectivity, $\omega_0 = \sqrt{b_0/b_2} \simeq \omega_i$ and

$$Q_m = \frac{\sqrt{b_0 b_2}}{b_1} = \frac{\omega_0}{2\sigma_i} \simeq \frac{\omega_i}{2\sigma_i} \qquad (4.7)$$

The latter is a measure of the degree of resonance. Thus, $Q_m < 0.5$ corresponds to a non-resonant (passive-RC) response, $Q_m = 0.5$ to a critically-damped situation, while $Q_m = \infty$ represents a zero-damped or oscillatory system. So convenient is this definition for the popular RC-active biquadratic sections, that it is henceforward referred to as 'Q' for the remainder of the text. For the

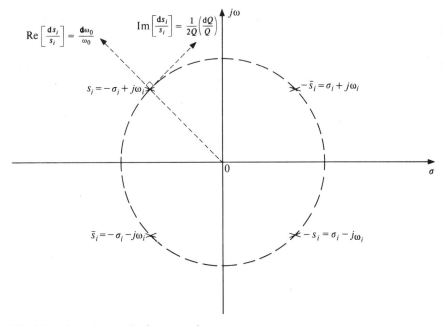

Fig. 4.3. Poles in the complex-frequency plane

bandpass response of Fig. 4.1, $N(s) = Hs$ and the centreband gain $K_0 = HQ/\omega_0$. When n such identical stages are cascaded, well-known[6] bandwidth analyses enable the overall Morris Q to be written as

$$Q_n = \frac{Q}{\sqrt{2^{1/n} - 1}} \qquad (4.8)$$

4.1.2 Rejective responses

Band-elimination filters rely on the use of notch-response networks with a transmission zero at a finite frequency (Ω_0, as normalized in Fig. 4.4). In such cases, half-power-point definitions have little significance as it is the notch rather than the response-shoulder characteristics which are of interest. Attempts[7] to invoke an inverted-response half-power definition (see Table 4.1) can lead to inconsistencies with values expected on a fundamental basis, and its full utility requires normalization with respect to a factor $1 - |T(j\Omega_0)|$. Alternatively, various response-slope measures can be considered around the null, the most universal being the normalized maximum gain slope.[2] This may be applied to null and minimum-transmission, terminated and non-terminated, lumped and distributed networks. Its shortcoming lies in the awkwardness of evaluation, especially when applied to measured responses. For the latter, it may be necessary to resort to empirical approximants. The definitions listed in

SELECTIVITY, SENSITIVITY, AND STABILITY

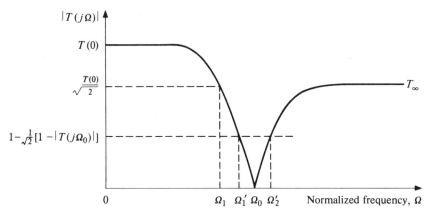

Fig. 4.4. Rejective frequency response

Table 4.1 'Rejectivity' definitions

Definition	Q expression	Comments
Inverted half-power	$\dfrac{\Omega_0}{\Omega_2' - \Omega_1'}$	Requires normalization for consistency
Phase slope	$\dfrac{1}{2}\dfrac{d}{d\Omega}[\arg T(j\Omega)]\big|_{\Omega=1}$	Only valid for high Q
Gain slope	$\dfrac{1}{2}\dfrac{d}{d\Omega}[|T(j\Omega)|]\big|_{\Omega=1}$	Not applicable to minimum-transmission and asymmetrical responses
Maximum gain slope	$\dfrac{1}{2\sqrt{T(0)T(\infty)}} \cdot \dfrac{d}{d\Omega}|T(j\Omega)|\big|_{\max Q \geqslant 1}$	Completely general but difficult to evaluate

Table 4.1 are illustrated by Fig. 4.4 in terms of normalized frequency with $\Omega_0 = 1$. For a null-adjusted second-order RC network, Eq. (3.34) gives

$$T(s) = \frac{(s/\omega_0)^2 + 1}{(s/\omega_0)^2 + (1/Q)(s/\omega_0) + 1} \tag{4.9}$$

and the definitions are compatible, although fundamental energy and gain-slope considerations indicate that selectivity would be more appropriately defined as twice that in Eq. (4.7). Should the network be out of adjustment, the null degenerates into a point of minimum transmission with a finite notch depth which also becomes an important parameter in practical situations.

4.1.3 Selectivity factor

High-order filter systems are not amenable to the criteria discussed above and it is usual to relate their selectivity to the rate at which the amplitude response

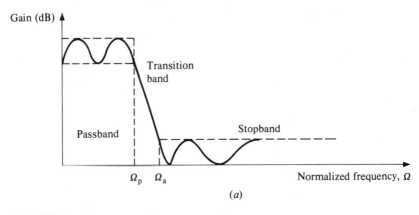

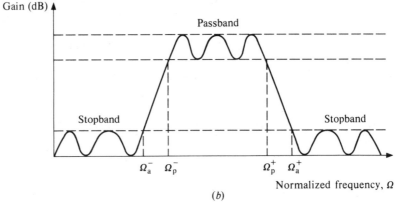

Fig. 4.5. (*a*) Lowpass gain-frequency response; (*b*) bandpass gain-frequency response

in decibels is cut off in the transition from the passband to stopband or vice versa. With reference to Fig. 4.5(*a*), the selectivity factor is defined[3] as

$$k = \sin^{-1} \theta = \frac{\Omega_p}{\Omega_a} \qquad (4.10)$$

where θ is the modular angle of an elliptic filter (see Chapter 2). In the case of a bandpass response characteristic, as in Fig. 4.5(*b*),

$$k_1 = \sin^{-1} \theta_1 = \frac{\Omega_p^+ - \Omega_p^-}{\Omega_a^+ - \Omega_a^-}$$

and for bandstop, the inverse. This, however, says nothing of the fractional bandwidth of the filter, i.e., $\sqrt{\Omega_p^+ \Omega_p^-}/(\Omega_p^+ - \Omega_p^-)$, which can also be regarded as a measure of selectivity. Defining

$$k_2 = \ln\left(1 + \frac{\sqrt{\Omega_p^+ \Omega_p^-}}{\Omega_p^+ - \Omega_p^-}\right)$$

then the overall filter selectivity has been defined[8] as $\sqrt{k_1^2 + k_2^2}$ and considerable insight into the performances of alternative filter realizations can be gained by using k_1 and k_2 as orthogonal axes of a 'selectivity plane'.

4.2 Single-parameter sensitivity

The selectivity discussed above must not only be attained but also maintained.[9] This can be a special problem in active networks where selectivity is achieved by using non-reciprocal elements to counterbalance losses in the passive components. Environmental variation in operating conditions can cause changes in the active and passive parameters, thereby giving rise to pole and zero displacements manifested by irregularities in filter performance (for a full explanation refer to Sec. 9.5). Such an occurrence is usually measured by the sensitivity defined[1] in its response/cause ratio sense. The diversity of sensitivity definitions for different circumstances is detailed in the ensuing subsections. These may be used as bases of comparison for the different network realizations to be discussed in later chapters. It is convenient initially to develop sensitivity analysis with respect to a single critical parameter as it is more amenable to minimization. But first, network stability needs to be ascertained.

4.2.1 Stability

Once more, multiplicity of definitions is evident but, for the present purpose, system stability is taken[1] as 'the property that bounded input perturbations result in bounded output peturbations.' In other words, it expresses the inability of the natural behaviour of a network to grow without limit (or, in practice, until it is limited by non-linearities and power-handling capability). Consideration of the realizability conditions of Sec. 3.3 with reference to the complex-frequency plane of Fig. 4.3 indicates that a linear network may be classified[10] in the categories:

(a) *strictly stable*, i.e., the network function has no poles either in the right half-plane or on the $j\omega$ axis;
(b) *marginally stable*, i.e., the network function has any single pair of conjugate poles on the $j\omega$ axis and all others in the left half-plane;

(c) *unstable*, i.e., the network function has poles either in the right half-plane or non-simple on the $j\omega$ axis. In the time domain, this implies that the impulse response has terms increasing with time.

For the case when the function is an impedance, the presence of poles exclusively in the left half-plane is said to denote *open-circuit stability* while, when the same is true for admittance functions, it is said to denote *short-circuit stability*. The two conditions need not be satisfied simultaneously, as a momentary consideration of the allpass function (Eq. (2.21)) confirms. In the usual practical situation when the network is terminated by the impedances of source and load, it is said to be *absolutely stable* if it is *strictly stable* for all possible passive terminations. If a passive terminating immittance exists for which the total port immittance (i.e., driving point plus termination) has at least one pole in the right half-plane or on the $j\omega$ axis, then the network is said to be *potentially unstable*. Hence it is possible, for example, for source nonidealities to restore stability.

The conditional stability of a linear two-port network is determined by the *stability invariant factor* derivable[10] from Llewellyn's conditions. Such an approach is beyond the scope of this book, the aim in practical filter design being to achieve a given response while assuring stability by more pragmatical inspection of functions. For the purposes of the remainder of this chapter, it is sufficient to appreciate that instability is an inherent property of an active network brought about by the introduction of feedback. While this enables high selectivity to be obtained by operating on the verge of oscillation ($\sigma_i \to 0$ in Eq. (4.7)), the consequent position of the transfer-function poles close to the $j\omega$ axis renders the performance very susceptible to component variations. Hence, in such cases, the sensitivity deteriorates with increasing Q, possibly necessitating redesign to ensure stability.

4.2.2 Differential sensitivity

The *classical* or *relative* measure of sensitivity based on an inversion of Bode's definition[11] is usually accepted as

$$S_x^T = \frac{\Delta T/T}{\Delta x/x}\bigg|_{\Delta x \to 0} = \frac{x}{T}\frac{\partial T}{\partial x} = \frac{\partial(\ln T)}{\partial(\ln x)} \qquad (4.11)$$

representing the effect of a differential change in some element x to the resulting differential change in a network performance parameter T. The transfer-function symbol is adopted for convenience here as it is usually the quantity of interest, but the expressions are equally valid for immittance functions. Simple application of Eq. (4.11) yields some interesting identities[12] summarized in Table 4.2. A wide range of sensitivity functions is covered by the first three entries while the next three, which can be useful for sensitivity

SELECTIVITY, SENSITIVITY, AND STABILITY

Table 4.2 Classical sensitivity identities

1.	$S_{x^n}^{aT_1^{m_1}T_2^{m_2}} = \dfrac{1}{n}(m_1 S_x^{T_1} + m_2 S_x^{T_2})$
2.	$S_x^{T_1+T_2} = \dfrac{(T_1 S_x^{T_1} + T_2 S_x^{T_2})}{(T_1 + T_2)}$
3.	$S_x^T = S_\zeta^T \cdot S_x^\zeta$
4.	$S_x^{\ln T} = \dfrac{1}{\ln T} S_x^T$
5.	$S_x^{\sinh T} = T \cdot \coth T \cdot S_x^T$
6.	$S_x^{\coth T} = T \cdot \tanh T \cdot S_x^T$
7.	$S_x^T = \text{Re}[S_x^T]$
8.	$S_x^{\arg T} = \text{Im}[S_x^T]$
9.	$S_{u/v}^T = S_u^T - S_v^T$

T, T_1, T_2 are network functions of parameter x; ζ, u, v are single-valued functions of x; m_1, m_2, n are real numbers; a is a constant.

assessment of distributed networks represented by irrational hyperbolic functions (see Chapter 3), can obviously be extended to sinusoidal functions by introduction of the operator 'j'. For a network function $T(s)$ with sinusoidal excitation,[13]

$$S_x^{T(j\omega)} = \frac{d[\ln|T(j\omega)|e^{j\phi(\omega)}]}{d[\ln x]}$$

$$= S_x^{|T(j\omega)|} + jx\frac{d\phi(\omega)}{dx} \qquad (4.12)$$

$$= S_x^{|T(j\omega)|} + j\phi(\omega) S_x^{\phi(\omega)}$$

where

$$\phi(\omega) = \arg T(j\omega)$$

$$S_x^{|T(j\omega)|} = \text{gain sensitivity}^{10}$$

and

$$x\frac{d\phi(\omega)}{dx} = \text{phase sensitivity}$$

Hence entries 7 and 8 in Table 4.2. From Eq. (4.12), it is clear that $|S_x^T| \geqslant |S_x^T|$. These sensitivity measures are useful for computing the effect of incremental parameter variations on the frequency response over a wide band of interest, but there are alternative definitions which are more meaningful for a particular performance characteristic. Moreover, the difficulties of evaluation already discussed in connection with the gain slopes of Table 4.1 are also in evidence here for all but the simplest of network configurations.

The classical definition may be used to investigate the effect of feedback on sensitivity. By application of the signal-flow techniques of Appendix 4, $S_x^{T(s)}$ can be found[10,13] as the reciprocal of the loop gain or return difference $F(s)$ of a feedback system. This implies that the transfer-function sensitivity is reduced by heavy feedback.

If x is any parameter of a lumped network, then it is known[11] that the coefficients of both numerator and denominator of the network function $T(s, x)$ depend linearly on x. From this fact, further sensitivity relations may be enounced. The absolute *root*[14] or *pole-position*[13] *sensitivity* is defined as

$$s_x^{s_i} = \frac{\mathrm{d}s_i}{\mathrm{d}x} \tag{4.13}$$

where s_i is a pole or zero of $T(s, x)$ as in Fig. 4.3 and is equal to the residue of the corresponding return-difference-function pole. In cases such as high-Q bandpass filters, where the proximity of dominant poles and zeros to the $j\omega$ axis is of interest, their displacement by changes in network elements is more appropriately estimated[10] by the semi-relative *pole-zero sensitivity*

$$S_x^{s_i} \simeq x \frac{\mathrm{d}s_i}{\mathrm{d}x}\bigg|_{s=s_i} \tag{4.14}$$

which is, in general, a complex number. It is actually[15] the residue at a pole or zero of the partial-fraction expansion of $S_x^{T(s)}$ giving a measure of the change in location of a pole or zero caused by a relative change in network element x. Since the pole and zero sensitivities define the root sensitivities of the denominator and numerator polynomials, respectively, then the coefficient sensitivity of the latter may be invoked for convenience. For a polynomial

$$\delta(s, x) = \sum_{j=0}^{n} \delta_j s^j \tag{4.15}$$

the *coefficient sensitivity* may be defined[10] as

$$S_x^{\delta_j} = \frac{\mathrm{d}(\ln \delta_j)}{\mathrm{d}(\ln x)} \tag{4.16}$$

which is, again, complex but not measurable. As expected, a simple relation exists between the pole-zero and coefficient sensitivities expressed by

$$S_x^{s_i} = -\frac{(s-s_i)}{\delta(s)}\bigg|_{s=s_i} \left[\sum_{j=0}^{n} \delta_j S_x^{\delta_j} s_i^j\right] \quad (4.17)$$

where s is a simple zero of $\delta(s)$ which moves as a result of an incremental increase in each coefficient. This gives rise to a sensitivity matrix representing the interrelationship for the entire polynomial. Further, the column matrix $[\Delta\delta^{(n)}]$ representing incremental changes in n coefficients can be related[16] to column matrix $[\Delta x^{(n)}]$ representing elemental increments by the sensitivity matrix $[S]$ for first-order effects:

$$[\Delta\delta^{(n)}] = [S][\Delta x^{(n)}] \quad (4.18)$$

4.2.3 Sensitivity minimization

While a totally insensitive network is unrealistic, it is often possible to minimize sensitivity with respect to a single component. This suggests[12] differentiating the sensitivity function by a pertinent parameter so that, in general,

$$\frac{d}{dx_2}(S_{x_1}^T) = \frac{1}{x_2} S_{x_1}^T [S_{x_1}^{x_2} + S_{x_2}^{dT/dx_1} - S_{x_2}^T] \quad (4.19)$$

from which the minima can be determined after distinguishing maxima and points of inflexion. When $x_2 = x_1 = x$, it indicates that the unity-sensitivity levels are significant, but few conclusions can be drawn unless the character of $T(x)$ is specified.

Another approach has been mentioned in Sec. 4.2.1, i.e., by increasing the loop gain of the system. This is the key to the low-sensitivity performance of many realizations to be discussed in Chapter 7 where the high gain of operational amplifiers is employed to increase the loop gain. However, the latter is limited in rate of cut-off by stability considerations which necessitate a wide-band network for high loop gain. Methods peculiar to the minimization of each of the sensitivity measures in Sec. 4.2.1 can be developed.[10] For example, the sensitivity of a pole or zero $s_i = \sigma_i + j\omega_i$ (see Fig. 4.3) can be expressed from Eq. (4.14) as

$$S_x^{s_i} = \frac{\Delta\sigma_i + j\Delta\omega_i}{\Delta x/x} = |S_x^{s_i}| \angle \tan^{-1}\left(\frac{\Delta\omega_i}{\Delta\sigma_i}\right) \quad (4.20)$$

so that, not only must the modulus be small, but stability requires that the argument must approach $\pm 90°$ in order that displacement of the root in the complex plane should be parallel to rather than towards the $j\omega$ axis.

Optimal sensitivity decompositions As is to be seen in later chapters, RC-active synthesis often depends on the partitioning of network functions $P(s)$ into realizable subnetwork immittance or transfer functions $P_a(s)$ and $P_b(s)$, each with negative-real zeros. Hence

$$P(s) = P_a(s) \pm kP_b(s) \qquad (4.21)$$

where k is a constant determined by the active element. To this end, a polynomial $D(s)$ of the function must be decomposed[17] into the difference or sum of two other polynomials $D_a(s)$ and $D_b(s)$ each with negative-real zeros. Developing the network function as a quotient,

$$P(s) = \frac{D(s)}{Q(s)} = \frac{M_a(s)}{Q_a(s)} \pm k\frac{M_b(s)}{Q_b(s)} \qquad (4.22)$$

giving

$$D(s) = D_a(s) \pm kD_b(s) \qquad (4.23)$$

where $Q(s) = Q_a(s)\,Q_b(s)$ is a divisor polynomial with all-distinct negative-real zeros. The problem is to choose the decomposition in such a manner that the sensitivity to the active parameter is minimal. The two cases in Eq. (4.23) are next considered in turn, though it should be remembered that the negative sign does not necessarily imply a difference effect nor the positive sign a summation as mixtures of driving-point and transfer immittance functions for unspecified $P_a(s)$ and $P_b(s)$ can invert the sequence.

Difference decomposition For active networks presenting both $P_a(s)$ and $P_b(s)$ as RC driving-point immittances separated by a negative sign in Eq. (4.21), several differential sensitivities are simultaneously minimized by means of a unique decomposition.[18] This may be deduced[19] algebraically by assuming that $D(s)$ does not have negative-real roots and hence employing the analytic continuation technique of Chapter 2,

$$D(s^2) = \pm H(s)\,H(-s) \qquad (4.24)$$

where $H(s)$ is a Hurwitz polynomial mentioned in Chapter 3. The positive sign refers to even order and the negative sign to odd order. Then, decomposing $H(s)$ into the sum of its even part $A(s^2)$ and odd part $sB(s^2)$ gives

$$D(s^2) = \pm \{A(s^2) + sB(s^2)\}\{A(s^2) - sB(s^2)\}$$
$$= \pm \{A^2(s^2) - s^2B^2(s^2)\}$$

which, on replacing s^2 by s, using the transformation of Sec. 3.2.1, yields the *Horowitz decomposition*

$$D(s) = \pm \{A^2(s) - sB^2(s)\} \qquad (4.25)$$

$A^2(s)$ and $B^2(s)$ can be identified with terms in Eq. (4.23), non-unity k being absorbed in $M_b(s)$ and/or $Q_a(s)$ of Eq. (4.22). Originally established on the basis of coefficient sensitivity using a root-locus approach, the even-order case can be expressed in the form

$$D(s) = \sum_{i=1}^{n} a_i s^i = \prod_{i=1}^{n/2} (s + \alpha_i)^2 - ks \prod_{i=1}^{n/2-1} (s + \beta_i)^2 \qquad (4.26)$$

This unique form with repeated roots α_i and β_i on the negative-real axis also minimizes pole-zero sensitivity besides providing optimal sensitivity to passive elements. Its application is particularly prevalent in realizations employing the negative-impedance converter (see Chapter 5) for which k becomes the conversion constant. Any negative-real zeros in $D(s)$ can be isolated by factorization and realized by a separate cascade section. However, use of difference effects as a means of achieving high Q is not generally recommended as this involves a small sensitive difference of large quantities.

Summation decomposition In Eq. (4.21), when $P_a(s)$ is an RC driving-point admittance and $P_b(s)$ an RC driving-point impedance (or an RL driving-point admittance) separated by a positive sign, then, of the many possible forms of decomposition of polynomial $D(s)$, that given by

$$D(s) = \prod_{i=1}^{m} (s + s_i)(s + \bar{s}_i)$$

$$= A^2(s) + B^2(s)$$

$$= \prod_{i=1}^{m} (s + \alpha_{2i-1})^2 + k \prod_{i=1}^{l} (s + \beta_{2i-1})^2 \qquad (4.27)$$

where s_i is a complex root and $\bar{s}_i$ its conjugate, α_{2i-1} and β_{2i-1} are negative-real roots, and $l = m$ or $m - 1$, is known[20] to be optimal with respect to pole-zero sensitivity. That is, when it exists at all subject to the angle criterion

$$\sum_{i=1}^{m} \arg s_i \leqslant \frac{\pi}{2} \qquad (4.28)$$

which can be thought of[17] as ensuring that the overall phase effect is restricted to the left-half plane of Fig. 4.3. This condition obviously accommodates second-order functions, but is more limiting for higher orders. Unlike the difference decomposition, there is no unique summation form since, when the degrees of $A(s)$ and $B(s)$ are equal (i.e., $l = m$), there are always more unknown quantities than known numbers in the solution of Eq. (4.27). Only the alternative case with $l = m - 1$ is unique. Hence there is considerable flexibility in partitioning the associated *RC-RL* network functions and the degree of freedom can be exploited to optimize other aspects of *RC*-gyrator realization (see Chapter 8) by computer-aided design techniques. The pole-zero sensitivity to k is found to be simultaneously minimized for any decomposition of this type and, subject to the severe limits to applicability set by Eq. (4.28), is lower than for any difference decomposition. Formation of decompositions is clarified by the numerical example below.

Example 4.1 Determine difference and summation decompositions for the polynomial[20] given by

$$D(s) = (s^2 + 2s + 2)(s^2 + \sqrt{3}s + 1)$$

and compare the corresponding zero sensitivities.

SOLUTION $D(s)$ is a fourth-order polynomial, so putting $n = 4$ in Eq. (4.26) and solving simultaneously for α_i and β_i terms of coefficients a_i gives the difference decomposition

$$D(s) = (s + 0.63)^2(s + 2.25)^2 - 2.04s(s + 1.14)$$

Factorizing,

$$D(s) = \prod_{i=1}^{2} (s + s_i)(s + \bar{s}_i)$$

$$= (s + 1 - j)(s + 1 + j)(s + \sqrt{3}/2 - j\tfrac{1}{2})(s + \sqrt{3}/2 + j\tfrac{1}{2})$$

and

$$\frac{dD(s)}{ds_i} = \frac{D(s)}{s + s_i}$$

Then, by Eq. (4.14), the corresponding modulus of zero sensitivity of $-s_i$ to k is given by

$$S_k^{-s_i} \simeq \left| k \frac{ds_i}{dk} \right|_{s=-s_i} = \left| k \frac{dD(s)}{dk} \frac{s + s_i}{D(s)} \right|_{s=-s_i}$$

From Eq. (4.23),

$$|kD_b(-s_i)| = |D_a(-s_i)|$$

Therefore

$$S_k^{-s_i} \simeq \left| \frac{D_a(-s_i)(s + s_i)}{D(-s_i)} \right|$$

and

$$S_k^{-1+j} = \left|\frac{(0.63 - 1 + j)^2(2.25 - 1 + j)^2}{(-1 + j + 1 + j)(-1 + j + \sqrt{3}/2 - j\tfrac{1}{2})(-1 + j + \sqrt{3}/2 + j\tfrac{1}{2})}\right|$$

$$= 1.87$$

Similarly,

$$S_k^{-\sqrt{3}/2 + j/2} = 0.85$$

Substituting from the factorized form into Eq. (4.28) gives

$$\left|\sum_{i=1}^{2} \arg s_i\right| = \frac{\pi}{4} + \frac{\pi}{6} = \frac{5\pi}{12}$$

which satisfies the angle criterion and allows a summation decomposition. By simultaneous solution or otherwise[16] for $m = 2$ and $l = 1$, Eq. (4.27) becomes

$$D(s) = (s + 0.226)^2(s + 1.64)^2 + 2.25(s + 0.91)^2$$

and the associated zero sensitivities this time are

$$S_k^{-1+j} = \left|\frac{(0.226 - 1 + j)^2(1.64 - 1 + j)^2}{(-1 + j + 1 + j)(-1 + j + \sqrt{3}/2 - j\tfrac{1}{2})(-1 + j + \sqrt{3}/2 + j\tfrac{1}{2})}\right|$$

$$= 1.45$$

and

$$S_k^{-\sqrt{3}/2 + j/2} = 0.714$$

Any other summation decomposition of this form suitable for *RC-RL* realization would give the same results, which are seen to be lower than for the corresponding difference-decomposition sensitivities for *RC-NIC* realizations.

Prescribed-sensitivity decompositions There are limits to the extent of minimization possible by difference and summation decompositions and really low sensitivities may demand degeneration of components. If, however, the *RC*-immittance function constraints are relaxed then it is possible[21] to achieve arbitrarily prescribed sensitivities. One method is to control the pole sensitivity by choice of the feedback loop or 'phantom' zeros such that changes in the active parameter do not give rise to more than definite displacements of the transfer-function poles. In terms of Eq. (4.21), function $P_b(s)$ is selected to define the phantom zeros for a prescribed pole sensitivity, then parameter k is controlled so that $P_a(s)$ has distinct negative-real zeros. Although S_k^p may be made as small as desired in this manner, there is the remaining difficulty of realizing $P_b(s)$ with low sensitivities to variations in other components.

4.2.4 Sensitivity of second-order systems

Since polynomial coefficients have a marked effect on the poles or zeros of a high-order function, it is advisable to control the roots rather than the coefficients. Restriction to first and second orders effectively allows this, suggesting that high-order filter transfer functions be factorized and realized as a non-interacting buffered cascade. Discussion of which categories of filter benefit by such realization and of choice of factorization is deferred until Chapter 9. It is well established that the second-order cascade approach has sensitivity advantages as compared with single-amplifier realization *en bloc*. This can be demonstrated[22] by considering the realization of a fourth-order factorizable transfer function: first as a product of two functions

$$\frac{N_1(s)}{s^2 + 2\sigma_1 s + \omega_1^2} \quad \text{and} \quad \frac{N_2(s)}{s^2 + 2\sigma_2 s + \omega_2^2}$$

then as an entirety. Each second-order section has optimum difference decomposition of the denominator with respect to active element gains k_i ($i = 1, 2$) as

$$(s - s_i)(s - \bar{s}_i) = s^2 + 2\sigma_i s + \omega_i^2 = (s + \alpha_i)^2 - k_i \beta_0 s \quad (4.29)$$

where it is assumed that the pole Q's are high, i.e., $\sigma_i \simeq 0$. Similarly, for the single-stage realization with respect to active-element gain k, the decomposition minimizing root sensitivities is

$$(s - s_1)(s - \bar{s}_1)(s - s_2)(s - \bar{s}_2) = (s + \alpha_2)^2 (s + \alpha_3)^2 - k\beta_1 s(s + \beta_2)^2 \quad (4.30)$$

Equating coefficients in Eqs. (4.29) and (4.30) allows identification of α's and β's and, for pole $s_1 = -\sigma_1 - j\omega_1$, gives pole-zero sensitivities

$$S_{k_1}^{s_1} = \frac{k_1 \beta_0 s (s + \sigma_1 + j\omega_1)}{s^2 + 2\sigma_1 s + \omega_1^2} \bigg|_{s=-\sigma_1-j\omega_1} \simeq (2Q - 1)\sigma_1 \quad (4.31)$$

$$S_{k_1}^{s_1} = \frac{k\beta_1 s(s + \beta_2)^2}{(s^2 + 2\sigma_2 s + \omega_2^2)(s + \sigma_1 - j\omega_1)} \bigg|_{s=-\sigma_1-j\omega_1} \simeq 4Q \left| \frac{\omega_2 + \omega_1}{\omega_2 - \omega_1} \right| \sigma_1 \quad (4.32)$$

assuming close proximity of poles ($\omega_1 \simeq \omega_2$) and where, by Eq. (4.7), $Q = \omega_1/2\sigma_1$. These results indicate the sensitivity enhancement achieved (albeit under special conditions) at the expense of extra active elements in a cascade realization.

SELECTIVITY, SENSITIVITY, AND STABILITY

On account of the above advantages and their widespread use in filter systems, it is convenient to define sensitivity measures peculiar to second-order sections as described by the transfer function of Eq. (4.6). The quantities Q and ω_0 are of practical significance[22] because of their ready measurement. The expression for Q associated with Eq. (4.7) illustrates the desirability of pole displacements parallel to the $j\omega$ rather than the σ axis in the complex plane of Fig. 4.3 as Q is particularly sensitive to small changes in σ. The *Q sensitivity*, giving a measure of how response shape and stability margin are affected by incremental changes in element x, can be defined[10] as

$$S_x^Q = \frac{x}{Q} \frac{dQ}{dx} \qquad (4.33)$$

and the *frequency sensitivity*[23] as

$$S_x^{\omega_0} = \frac{x}{\omega_0} \frac{d\omega_0}{dx} \qquad (4.34)$$

Their relative importance in ensuring a specified filter response may be assessed by considering the classical sensitivity of a pole[9] from Eq. (4.14) with $\omega_0 = \sqrt{\sigma_i^2 + \omega_i^2}$ and $Q = \omega_0/2\sigma_i \simeq \omega_i/2\sigma_i$ for steady-state sinusoidal excitation as

$$S_x^{s_i} = \frac{1}{\sigma_i + j\omega_i} S_x^{s_i} = \frac{d\sigma_i + j d\omega_i}{\sigma_i + j\omega_i} \cdot \frac{x}{dx}$$

$$= \left[\left(\frac{\sigma_i d\sigma_i + \omega_i d\omega_i}{\sigma_i^2 + \omega_i^2} \right) + j \left(\frac{\sigma_i d\omega_i - \omega_i d\sigma_i}{\sigma_i^2 + \omega_i^2} \right) \right] \frac{x}{dx}$$

$$= S_x^{\omega_0} + j \frac{1}{2Q} S_x^Q \qquad (4.35)$$

With reference to Fig. 4.3, this indicates that the real part represents radial motion of the pole in the complex plane, which changes the resonant frequency but does not affect the Q. For any absolutely stable passive network, the imaginary part is inversely proportional to Q and hence the effect of Q sensitivity, which is independent of selectivity, cannot be excessive. On the other hand, an active feedback network (see Sec. 7.1) made selective by operating on the verge of oscillation becomes more sensitive as Q is increased.

The nature of the result[24] in Eq. (4.35) may be further revealed[25] by developing transfer-function sensitivities for the biquadratic form of Eq. (2.1) with logarithmic gain

$$\gamma(\omega) = 20 \log_{10} |T(j\omega)| \quad \text{dB}$$

$$= 20 \log_{10} |(a_0 - a_2 \omega^2) + ja_1\omega| - 20 \log_{10} |(b_0 - b_2\omega^2) + jb_1\omega| \quad \text{dB}$$

where a_0, a_1, a_2 are the coefficients of the numerator polynomial $N(\omega)$ and b_0, b_1, b_2 are the coefficients of the denominator polynomial $D(\omega)$. Since the quadratic function

$$\delta(s) = \delta_2 s^2 + \delta_1 s + \delta_0$$

has $\delta_2 = 1$, by Eq. (4.6), $\delta_1 = \omega_0/Q$, and $\delta_0 = \omega_0^2$, then

$$S_{\omega_0}^y = S_{\delta_1}^y S_{\omega_1}^{\delta_1} + S_{\delta_0}^y S_{\omega_0}^{\delta_0} = S_{\delta_1}^y + 2S_{\delta_0}^y$$

and

$$S_Q^y = -S_{\delta_1}^y$$

Hence

$$S_{\omega_0}^y/S_Q^y = -1 - 2S_{\omega_0}^y/S_{\delta_1}^y \qquad (4.36)$$
$$= -1 - 2\delta_0(\delta_0 - \delta_2\omega^2)/(\delta_1\omega)^2$$
$$= -1 - 2Q^2(\omega_0^2/\omega^2 - 1) \qquad (4.37)$$

At the half-power frequencies,

$$|T(j\omega)| = 1/\sqrt{2}$$

and

$$\omega/\omega_0 = 1 \pm 1/2Q \qquad (4.38)$$

Substituting Eq. (4.38) in (4.37) and with Table 4.2 gives

$$S_x^Q/S_x^{\omega_0} = S_{\omega_0}^y/S_Q^y = \pm 2Q \qquad (4.39)$$

implying that for the same change of gain in decibels, the deviation in selectivity can be $2Q$ times greater than that in ω_0. This not only confirms Eq. (4.35) but has, *en passant*, emphasized in Eq. (4.37) the necessity of considering frequency dependence. Closer examination of Eq. (4.37) shows that, within the bounds of Eq. (4.38), it is of order Q while far outside it is of the order Q^2. Hence the dominance of frequency sensitivity is less marked within the passband than at frequencies far away from resonance. Such facts should be borne in mind when comparing the sensitivity of performances of different filters in later chapters.

Sensitivity minimization with respect to gain of active element In order to investigate how a second-order function should be decomposed for synthesis so as to minimize the more critical frequency-sensitivity criterion $S_A^{\omega_0}$ with respect to the amplifier gain A (the usual active-element parameter, see Chapter 5), consider the denominator of Eq. (4.6):

$$D(s) = b_2 s^2 + b_1 s + b_0$$

SELECTIVITY, SENSITIVITY, AND STABILITY

In general, each coefficient is a function $f(A)$ of the gains of the amplifiers used in the active realization. Hence it can be expressed as

$$D(s) = \beta_2 f_2(A)s^2 + \beta_1 f_1(A)s + \beta_0 f_0(A) \tag{4.40}$$

where $\beta_0, \beta_1, \beta_2$ are constants determined by the passive elements. If A is the near-infinite open-loop gain of an operational amplifier, then it is analytically convenient to express $f_i(A)$ in terms of small reciprocals so that the gain-dependent functions may be written

$$f_i(A) = f_i(\infty)\left[1 + \sum_{j=1}^{N}\frac{n_{ij}}{A_j} + \sum_{k=1}^{N}\sum_{j=1}^{N}\frac{n_{ijk}}{A_j A_k} + \cdots\right] \tag{4.41}$$

where N is the total number of amplifiers employed and $n_{ij} \ldots$ is the coefficient of the remainder terms in $1/A_j \ldots$. Since $A_j \ldots$ are very large, then only the first-order terms need to be considered. Then, simplifying with the notation

$$\frac{n_i}{A} = \sum_{j=1}^{N}\frac{n_{ij}}{A_j} \tag{4.42}$$

enables Eq. (4.40) to be written as

$$D(s) = \left(1 + \frac{n_2}{A}\right)\left(\frac{s}{\omega_0}\right)^2 + \left(1 + \frac{n_1}{A}\right)\left(\frac{s}{\omega_0}\right) + \left(1 + \frac{n_0}{A}\right) \tag{4.43}$$

The perturbed centre frequency may then be determined as

$$\omega_k = \omega_0\left(1 + \frac{n_0}{A}\right)^{1/2}\left(1 + \frac{n_2}{A}\right)^{-1/2}$$

$$= \omega_0\left[1 + \frac{1}{2}\left(\frac{n_0}{A} - \frac{n_2}{A}\right) + \cdots\right] \tag{4.44}$$

Obviously, frequency sensitivity is minimized if the condition $n_0/A = n_2/A$ is satisfied. This is guaranteed if gain functions $f_0(A)$ and $f_2(A)$ in Eq. (4.40) are equal.

Assuming all gain functions to be positive, then it follows that the denominator polynomial having reduced sensitivity to amplifier gain, consistent with enhanced selectivity (i.e., a difference first-order term), has the general form[26] of decomposition

$$D(s) = (\beta_2 s^2 + \beta_1 s + \beta_0) + \beta_2 F_2(A)s^2 - \beta_1 F_1(A)s + \beta_0 F_2(A) \tag{4.45}$$

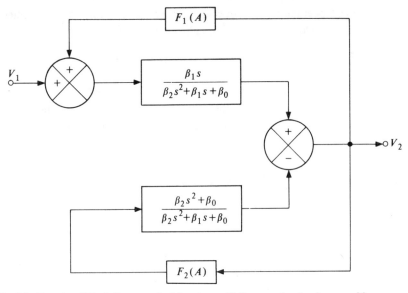

Fig. 4.6. Functional block diagram of minimum-sensitivity second-order decomposition

since

$$f_0(A) = f_2(A) = 1 + F_2(A)$$

This may be realized in block-diagram functional form as in Fig. 4.6. A number of special cases may be deduced from Eq. (4.45):

(a) a *bandpass positive-feedback* development characterized by $F_2(A) = 0$ giving rise to the family of circuits to be described in Chapter 6;
(b) a *notch-function negative-feedback* development characterized by $F_1(A) = 0$;
(c) an *allpass feedback* development characterized by $F_1(A) = F_2(A)$.

The synthesis problem is then one of realizing the passive functional blocks of Fig. 4.6. Their form is such as frequently to dictate use of the null-adjusted *RC* twin-T networks discussed in Chapter 3.

4.2.5 Sensitivity of higher-order systems

As is to be discussed in Chapter 9, realization by cascaded second-order blocks is not necessarily superior in sensitivity performance to direct formation of high-order filter functions. Furthermore, some basic second-order filter sections are fundamentally third-order circuits reduced by pole-zero cancellation (e.g., the *RC* twin-T network in Chapter 3). Consequently, an incremental sensitivity evaluation method for third- and higher-order systems is required.

SELECTIVITY, SENSITIVITY, AND STABILITY

Consider the third-order transfer function with $-\sigma$ as the cancelling real pole and zero:

$$T(s) = \frac{a_3 s^3 + a_2 s^2 + a_1 s + a_0}{s^3 + b_2 s^2 + b_1 s + b_0} = \frac{(s + \sigma)(d_2 s^2 + d_1 s + d_0)}{(s + \sigma)(s^2 + (\omega_0/Q)s + \omega_0^2)}$$

$$= \frac{(s + \sigma)(d_2 s^2 + d_1 s + d_0)}{(s + \sigma)(s + \sigma_i + j\omega_i)(s + \sigma_i - j\omega_i)} = \frac{(s + \sigma)(d_2 s^2 + d_1 s + d_0)}{(s + \sigma)(s^2 + 2\sigma_i s + \sigma_i^2 + \omega_i^2)}$$

(4.46)

where $d_2 = a_3$, etc. are introduced for symmetry. Equating denominator coefficients and identifying from Eqs. (4.6) and (4.7) gives

$$\begin{aligned} b_0 &= \sigma \omega_0^2 \\ b_1 &= \omega_0^2 + \omega_0 \sigma / Q \\ b_2 &= \sigma + \omega_0 / Q \end{aligned} \qquad (4.47)$$

while, using Eq. (4.14) and the δ coefficients of Sec. 4.2.4, pole-sensitivity parts with respect to parameter x are

$$S_x^{\sigma_i} = -\frac{\omega_0}{2Q}(S_x^{\omega_0} - S_x^Q)$$

and

$$S_x^{\omega_i} = \frac{\omega_0}{Q}\left(\sqrt{4Q^2 - 1}\, S_x^{\omega_0} + \frac{1}{\sqrt{4Q^2 - 1}} S_x^Q\right) \qquad (4.48)$$

By Eq. (4.15) and the identities of Table 4.2, the coefficient sensitivities may be expressed in matrix form:

$$\begin{bmatrix} S_x^{b_0} \\ S_x^{b_1} \\ S_x^{b_2} \end{bmatrix} = \begin{bmatrix} 2 & 0 & 1 \\ 1 + \dfrac{1}{1 + \dfrac{\sigma}{\omega_0 Q}} & \dfrac{-1}{1 + \dfrac{\omega_0 Q}{\sigma}} & \dfrac{1}{1 + \dfrac{\omega_0 Q}{\sigma}} \\ \dfrac{1}{1 + \dfrac{Q\sigma}{\omega_0}} & \dfrac{-1}{1 + \dfrac{Q\sigma}{\omega_0}} & \dfrac{1}{1 + \dfrac{\omega_0}{Q\sigma}} \end{bmatrix} \begin{bmatrix} S_x^{\omega_0} \\ S_x^Q \\ S_x^{\sigma} \end{bmatrix} \qquad (4.49)$$

Matrix inversion[27] then yields[23] the frequency, Q and real-pole sensitivities in terms of the more accessible coefficient sensitivities. For the normalized case with $\sigma = \omega_0 = 1$ and $b_0 = 1$, $b_1 = b_2 = 1 + 1/Q$, these become

$$S_x^{\omega_0} = \frac{Q-1}{2(2Q-1)} S_x^{b_0} + \frac{Q+1}{2(2Q-1)} (S_x^{b_1} - S_x^{b_2})$$

$$S_x^Q = \frac{Q+1}{2} (S_x^{b_0} - S_x^{b_1} - S_x^{b_2}) \qquad (4.50)$$

$$S_x^\sigma = \frac{Q}{2Q-1} S_x^{b_0} + \frac{Q+1}{2Q-1} (S_x^{b_2} - S_x^{b_1})$$

These may be readily evaluated with respect to each active and passive parameter of a network in order to identify the most sensitive components and compare different configurations.

The procedure may be extended to fourth-order filters by expressing the denominator of the transfer function as the product of quadratics. The matrix inversion is then circumvented by formulating[28] in terms of the relative pole sensitivity. Since the denominator polynomial

$$D(s) = \sum_{k=0}^{n} b_k s^k \qquad (4.51)$$

is zero at a root, then, differentiating,

$$\frac{d}{dx}\left(\sum_{k=0}^{n} b_k s_i^k\right) = 0$$

gives an expression for relative pole sensitivity in terms of coefficient sensitivities:

$$S_x^{s_i} = -\frac{\sum_{k=0}^{n} \beta_k s_i^k S_x^{\beta_k}}{\sum_{k=0}^{n} \beta_k s_i^k k} \qquad (4.52)$$

Then the frequency and Q sensitivities follow by application of Eq. (4.35) with

$$S_x^{\omega_0} \simeq \text{Re}(S_x^{s_i}) \quad \text{and} \quad S_x^Q \simeq 2Q \cdot \text{Im}(S_x^{s_i}) \qquad (4.53)$$

This involves a sequential algorithm of equations so that incremental sensitivity evaluation above fourth order remains intractable. Variation methods for determining pole-zero sensitivities of higher-order functions are available[29] if the design values of network elements are known but appear to have little practical utility.

4.2.6 Large-deviation sensitivity

The relative differential sensitivity of Eq. (4.11) may be approximated by the *incremental relative sensitivity* $\mathscr{S}_x^T$ giving the change ΔT in T resulting from a finite change Δx in parameter x. This is known[10] to be the reciprocal of the return difference evaluated at the perturbed value $x + \Delta x$. Although computational methods no longer necessitate approximation of classical sensitivity in this manner, the incremental representation is of value in assessing the effect of finite parameter changes. Since this is the practical case, it is interesting to consider to what extent the quantitative measure S_x^T reflects the actual change arising from Δx. When T is given as the ratio of two polynomials, they may be incremented[30] as follows:

$$T(s, x + \Delta x) = \frac{N(s, x + \Delta x)}{D(s, x + \Delta x)} = \frac{N + N'\Delta x}{D + D'\Delta x} \qquad (4.54)$$

where

$$N' = \frac{dN}{dx} \quad \text{and} \quad D' = \frac{dD}{dx}$$

Then, defining

$$\Delta T(s, x) \triangleq T(s, x + \Delta x) - T(s, x)$$

$$= \frac{N + N'\Delta x}{D + D'\Delta x} - \frac{N}{D}$$

$$= \frac{(N/D)'\Delta x}{1 + (D'/D)\Delta x} \qquad (4.55)$$

the fractional change in T resulting from a fractional element change $\Delta x/x$, i.e., the *relative variation*,[15] is

$$V_x^T = \frac{\Delta T}{T} = \mathscr{S}_x^T \frac{\Delta x}{x} \qquad (4.56)$$

Hence, from Eq. (4.55),

$$\mathscr{S}_x^T = \frac{\Delta T/T}{\Delta x/x} = \frac{S_x^T}{1 + S_x^D \Delta x/x} \qquad (4.57)$$

indicating the importance of the denominator sensitivity.

By Eq. (4.12), both the modulus and phase sensitivities must be zero to ensure zero differential sensitivity, thus rendering the component totally redundant. Equation (4.57) may be used[30] to derive an exact relationship with the modulus sensitivity which shows only the first- and second-order differentials of T to be significant.

4.3 Multiparameter sensitivity

One of the implications of integrated-circuit implementation of active filters is that the sensitivity of the active device is no longer of paramount importance. As will become apparent in Chapters 6 to 8, the passive elements can indeed cause a greater problem and their sensitivities become more critical as attempts are made to minimize active-device sensitivity. Simultaneous minimization of all sensitivities then constitutes a multiparameter problem.

4.3.1 Multiparameter variation analysis

If transmission function T is dependent on n parameters, i.e.,

$$T = T(x_1, x_2, \cdots, x_n) \tag{4.58}$$

then its variation with respect to all its elements can be obtained[29] from the total differential

$$dT = \frac{\partial T}{\partial x_1} dx_1 + \frac{\partial T}{\partial x_2} dx_2 + \cdots + \frac{\partial T}{\partial x_n} dx_n$$

as

$$\frac{dT}{T} = S^T_{x_1} \frac{dx_1}{x_1} + S^T_{x_2} \frac{dx_2}{x_2} + \cdots + S^T_{x_n} \frac{dx_n}{x_n}$$

Extending Eq. (4.56),

$$\frac{\Delta T}{T} = \sum_{i=1}^{n} V^T_{x_i} = \sum_{i=1}^{n} S^T_{x_i} \frac{dx_i}{x_i}$$

$$= d(\ln T) = \sum_{i=1}^{n} \frac{\partial [\ln T]}{\partial [\ln x_i]} d[\ln x_i] \tag{4.59}$$

neglecting second-order effects, i.e., assuming independent elements.

Expressing as a multiparameter sensitivity vector gives

$$\overline{S}^{T(s)}_x = \left[\frac{\partial (\ln T)}{\partial (\ln x_1)}, \frac{\partial (\ln T)}{\partial (\ln x_2)}, \cdots, \frac{\partial (\ln T)}{\partial (\ln x_n)} \right]$$

$$= [S^{T(s)}_{x_1}, S^{T(s)}_{x_2}, \cdots, S^{T(s)}_{x_n}]$$

SELECTIVITY, SENSITIVITY, AND STABILITY

which has real and imaginary parts yielding the multiparameter gain and phase sensitivities, respectively. A suitable design criterion[10] is to minimize the magnitude

$$|\overline{S_x^{T(s)}}| = \sqrt{\sum_{i=1}^{n} \left|\frac{\partial(\ln T)}{\partial(\ln x_i)}\right|^2} = \sqrt{[S_x^{T(j\omega)}]^* [S_x^{T(j\omega)}]^t} \qquad (4.60)$$

where $[\;]^t$ denotes the transpose of the sensitivity vector. This requires numerical optimization, particularly if the individual sensitivity parameters have different weightings.

4.3.2 Summed sensitivity

In order to derive the sum of the sensitivities of a function to network elements, it is necessary to consider the nature of a homogeneous function,[15,29] i.e., one for which scaling of elements $x_1, x_2, \cdots, x_n$ by an arbitrary parameter λ results in

$$T(\lambda x_1, \lambda x_2, \cdots, \lambda x_n) = \lambda^m T(x_1, x_2, \cdots, x_n) \qquad (4.61)$$

Differentiating with respect to λ, setting $\lambda = 1$, and dividing throughout by T gives

$$\frac{x_1}{T}\frac{\partial T}{\partial x_1} + \frac{x_2}{T}\frac{\partial T}{\partial x_2} + \cdots + \frac{x_n}{T}\frac{\partial T}{\partial x_n} = m \qquad (4.62)$$

which is Euler's formula representable as the summed sensitivities of T to network elements x_i as

$$\sum_{i=1}^{n} S_{x_i}^T = m \qquad (4.63)$$

Here, m is the order of homogeneity and is determined by dimensional considerations for a given type of network function so that

for T an impedance function, $m = +1$;
for T an admittance function, $m = -1$ and
for T a transfer function, $m = 0$.

Consider a general linear electrical network consisting of n_R resistors R_i, n_C capacitors C_i, n_L inductors L_i, and n_K active elements K_i described by function

$$T(aR_i, C_i/a, aL_i, K_i, s) = a^m T(R_i, C_i, L_i, K_i, s) \qquad (4.64)$$

after scaling impedances by factor a. Then, differentiating both sides with respect to a, setting $a = 1$, dividing throughout by T and expressing as summed sensitivities:

$$\sum_{i=1}^{n_R} S_{R_i}^T - \sum_{i=1}^{n_C} S_{C_i}^T + \sum_{i=1}^{n_L} S_{L_i}^T = m \qquad (4.65)$$

The negative sign for the capacitor sensitivity sum is a result of the negative power of x in the first identity of Table 4.2.

Now consider the effect of frequency scaling, with reactive components only factored by b, so that

$$T(R_i, bC_i, bL_i, K_i, s) = T(R_i, C_i, L_i, K_i, bs) \qquad (4.66)$$

Following the usual differentiation process then produces the invariant

$$\sum_{i=1}^{n_C} S_{C_i}^T + \sum_{i=1}^{n_L} S_{L_i}^T = S_s^T \qquad (4.67)$$

Restricting attention to RC-active networks, Eqs. (4.65) and (4.67) result[31] in

$$\sum_{i=1}^{n_R, n_C, n_K} S_{R_i, C_i, K_i}^T = \sum_{i=1}^{n_R} S_{R_i}^T + \sum_{i=1}^{n_C} S_{C_i}^T + \sum_{i=1}^{n_K} S_{K_i}^T = 2 S_s^T + \sum_{i=1}^{n_K} S_{K_i}^T + m \qquad (4.68)$$

So the passive constituent is only a function of $T(s)$ and not of the particular synthesis technique employed. It is interesting to observe that for a null-adjusted passive-RC transfer function, Eq. (4.68) is reminiscent of the gain-slope 'rejectivity' expression in Table 4.1, since then

$$\left|\frac{dT}{ds}\right|_{s=j\omega_0} = \frac{d|T|}{d\Omega}\bigg|_{\Omega=\omega/\omega_0=1}$$

This is suggestive of the inherent connection between selectivity and sensitivity, both being dependent on the energy storage in the network. In fact, the differential sensitivity function of any scattering parameter of a resistively terminated LC network has been shown[32] to be closely correlated to the energy stored in the reactive elements, being the geometric mean of the energies stored in the forward and reverse transmission. The sensitivity of a passive-reciprocal one-port immittance Z (or Y) to all reactive elements X_i can be expressed[33] as

$$\sum_{i=1}^{n_{LC}} S_{X_i}^Z = \frac{\omega(W_m - W_e)}{II^*|Z|} \qquad (4.69)$$

and to resistors as

$$\sum_{i=1}^{n_R} S_{R_i}^Z = \frac{R}{|Z|} \qquad (4.70)$$

where I is the port current, I^* is its conjugate, W_m and W_e are defined in Eq. (3.14), and n is the number of elements. The resemblance to the selectivity definition of Eqs. (4.2) and (4.3) is apparent. Unfortunately, for non-reciprocal networks, only lower and upper limits are available and the theory is not yet extended to active networks. However, these limits represent worst-case bounds and the sensitivity measure which emerges naturally from energy considerations is, from Eq. (4.69), the sum of the magnitudes of the sensitivities to individual elements, i.e., the *worst-case sensitivity*[34] criterion

$$W^T_{x_i} = \sum_{i=1}^{n} |S^T_{x_i}| \qquad (4.71)$$

As expected, this is pessimistic and, in fact,

$$W^{|T(j\omega)|}_{x_i} \geqslant 2|S^{|T(j\omega)|}_{\omega}|$$

implying that the lower bound is dependent only on the form of the transfer function. A further criterion frequently assumed is the sum of the magnitudes squared of the individual sensitivities, i.e., the *Schoeffler*[35] *sensitivity* derivable from Eq. (4.6) as

$$\phi^T_{x_i} = \sum_{i=1}^{n} |S^T_{x_i}|^2 \qquad (4.72)$$

If all n single-element sensitivities are equal in value, corresponding to

$$S^{|T(j\omega)|}_{x_i} = \frac{2}{n} S^{|T(j\omega)|}_{\omega}$$

then the lower bound on Eq. (4.72) is given by

$$\phi^{|T(j\omega)|}_{x_i} \geqslant \frac{4}{n}(S^{T(j\omega)}_{\omega})^2$$

which indicates that this measure can always be reduced arbitrarily by simply subdividing elements, i.e., by increasing n. Definitions (4.71) and (4.72) can be expressed in terms of the second-order Q and frequency sensitivities of Eqs. (4.33) and (4.34) together with the lower bound. On this basis, it is shown[36] that configurations with high $S^Q_{x_i}$ are not necessarily inferior to those with inherently low $S^Q_{x_i}$.

4.3.3 Multiparameter sensitivity minimization

Simultaneous minimization of the sensitivity of a network with respect to several parameters customarily invokes the theory of equivalent networks, i.e., the formation of one physically-realizable network from another such that the specified network response functions are invariant. In particular, the generation of continuously equivalent networks[35] whose elements differ by incremental amounts allows formulation of differential equations for element values and sensitivities which can be numerically optimized by means of Eq. (4.72). However, not all equivalent networks can be found[37] by this procedure and the optimization is restricted to a single frequency.

In order to minimize response deviations over a specified band of frequencies and to take account of component tolerances, computer-aided techniques can be used to full advantage by a statistical variability analysis. By virtue of integrated-circuit production processes, component tolerances are correlated with standard deviation $= \sigma$ (here assumed to be the same for all types of

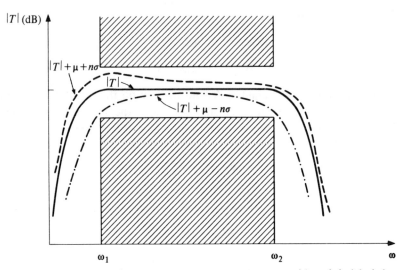

Fig. 4.7. Tolerance limits on specified amplitude-frequency response with statistical deviations

component). For a random deviation Δx_i in element x_i, the transfer function deviation is

$$\Delta T = T(j\omega, x + \Delta x) - T(j\omega, x) \qquad (4.73)$$

and over the specified frequency band from ω_1 to ω_2 in Fig. 4.7, the *statistical sensitivity function*,[24] is

$$M(x) = E\left[\int_{\omega_1}^{\omega_2} \left(\frac{\Delta T}{T}\right)^2 d\omega\right] \qquad (4.74)$$

where E is the expected value which essentially computes the average of the second moment of the function $\Delta T/T$. Assuming, initially, zero cross-correlation terms, then this can be expressed as

$$M(x) = \int_{\omega_1}^{\omega_1} \sum_{i=1}^{n} |S_{x_i}^T|^2 \sigma_{x_i}^2 d\omega \qquad (4.75)$$

where

$$\sigma_{x_i}^2 = E\left[\left(\frac{\Delta x_i}{x_i}\right)^2\right] \quad \text{for} \quad i = 1, 2, \ldots n,$$

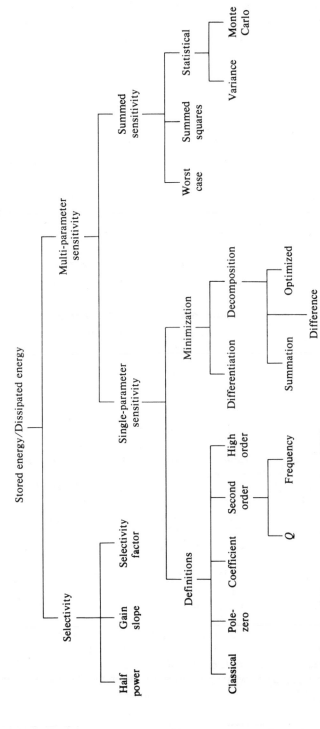

Fig. 4.8. Perspective chart of selectivity and sensitivity treatment

and n is the number of elements. It is obvious that the Schoeffler equation (Eq. (4.72)) is a special case of Eq. (4.75) for $\sigma_{x_i} = 1$.

If the mean change[38] μ in components as well as the variance is taken into account, then a computing algorithm can be developed to minimize $M(x)$ by successive iterations while maintaining the transfer function invariant. Figure 4.7 shows a range of distributions within the tolerance limits set on the response. Of course, in integrated-circuit form, strong correlations ($\rho_{x_1 x_2} \to 1$) exist between component values owing to their proximity and dependence on a limited number of process variables. Consequently, the associated cumbersome covariances need to be included in the evaluation of Eq. (4.74); Eq. (4.75) merely representing a lower bound in practice.

It is important to distinguish the above technique from that of minimization by means of Monte Carlo analysis since it employs analytic functions of a continuous variable ω and hence need only be tested at a single frequency to verify the response. By contrast, Monte Carlo methods[39,40] evaluate the network repeatedly for randomly generated sets of parameter values and tolerances. The resulting responses provide the statistical properties. It deals with inter-variable correlations and is not restricted to small deviations. On the other hand, it provides no information as to the relative importance of circuit parameters and is by its very nature extremely extravagant on computer time. In order to provide perspective, the definitions of this chapter are displayed in Fig. 4.8.

References

1. *IEEE Standard Dictionary of Electrical and Electronic Terms*, Wiley-Interscience, New York, 1972, p. 543.
2. P. Bowron and G. Wilson, 'Selectivity of lumped-passive RC notch filters', *Electron. Engng*, **46** (1), 46–51, 1974.
3. J. K. Skwirzinski, *Design Theory and Data for Electrical Filters*, Van Nostrand, Princeton, N.J., 1965, Chap. 2, p. 37.
4. D. R. Rhodes, 'Observable stored energies of electromagnetic systems', *J. Franklin Inst.* **302** (3), 225–237, 1976.
5. D. G. O. Morris, 'Q as a mathematical parameter', *Electron. Engng*, **26** (7), 306–308, 1954.
6. J. Millman and C. C. Halkias, *Integrated Electronics: Analog and Digital Circuits and Systems*, McGraw-Hill, Tokyo, 1972, Chap. 12, p. 386.
7. K. L. Su, 'Selectivity of notch filters using non-uniform RC lines', *Electron. Letters*, **1**, 204–205, 1965.
8. M. Novák, 'Synthese von Hochfrequenzfiltern hoher Selektivität, in *Eighth International Colloquium*, Ilmenau (City) Hochschule für Elektrotechnik, E. Germany, 1963, Part 1, pp. 31–36.
9. W. E. Newell, 'Selectivity and sensitivity in functional blocks', *Proc. IRE*, **50**, 2517, 1962.

10. S. K. Mitra, *Analysis and Synthesis of Linear Active Networks*, Wiley, New York, 1969.
11. H. W. Bode, *Network Analysis and Feedback Amplifier Design*, Van Nostrand, Princeton, N.J., 1945.
12. J. Gorski-Popiel, 'Classical sensitivity: a collection of formulae', *IEEE Trans*, **CT-10**, 300–301, 1963.
13. R. W. Newcomb, *Active Integrated Circuit Synthesis*, Prentice Hall, Englewood Cliffs, N.J., 1968.
14. F. F. Kuo, 'Pole-zero sensitivity in network functions', *IRE Trans.*, **CT-5**, 372–373, 1958.
15. G. S. Moschytz, *Linear Integrated Networks Fundamentals*, Van Nostrand Reinhold Co., New York, 1974.
16. L. P. Huelsman, *Theory and Design of Active RC Circuits*, McGraw-Hill, New York, 1968.
17. W. E. Heinlein and W. H. Holmes, *Active Filters for Integrated Circuits*, Prentice Hall, Englewood Cliffs, N.J.; R. Oldenbourg Verlag, Münich, 1974.
18. I. M. Horowitz, 'Optimisation of negative-impedance conversion methods of active *RC* synthesis', *IRE Trans.*, **CT-6**, 296–303, 1959.
19. R. E. Thomas, 'Polynomial decomposition in active RC network synthesis', *IRE Trans.*, **CT-8**, 270–274, 1961.
20. D. A. Calahan, 'Sensitivity minimisation in active *RC* synthesis', *IRE Trans*, **CT-9**, 38–42, 1962.
21. S. S. Haykin, *Synthesis of RC-active Filter Networks*, McGraw-Hill, London, 1969, Chap. 4.
22. G. S. Moschytz, 'A note on pole, frequency and Q sensitivity', *IEEE Journal*, **SC-6**, 267–268, 1971.
23. M. A. Soderstrand and S. K. Mitra, 'Sensitivity analysis of third-order filters', *Int. J. Electron.* **30** (3), 265–272, 1971.
24. A. L. Rosenblum and M. S. Ghausi, 'Multiparameter sensitivity in active *RC* networks', *IEEE Trans.*, **CT-18**, 592–599, 1971.
25. D. Hilberman, 'An approach to the sensitivity and statistical variability of biquadratic filters', *IEEE Trans.*, **CT-20**, 382–390, 1973.
26. G. Wilson, Y. Bedri, and P. Bowron, '*RC*-active networks with reduced sensitivity to amplifier gain-bandwidth product', *IEEE Trans.*, **CAS-21** (5), 618–626, 1974.
27. L. P. Huelsman, *Circuits, Matrices, and Linear Vector Spaces*, McGraw-Hill, New York, 1963.
28. S. C. DuttaRoy and D. K. Bhargava, 'Sensitivity of third and higher-order filters', *Int. J. Circuit Theory Applics*, **5**, 235–238, 1977.
29. C. Belove, 'The sensitivity function in variability analysis', *IEEE Trans.*, **R-15** (2), 70–76, 1966.
30. J. K. Fidler and C. Nightingale, 'Differential-incremental sensitivity relationships', *Electron. Letters*, **8** (5), 626–627, 1972.
31. A. G. J. Holt and M. R. Lee, 'Summed sensitivity of active *RC* networks', *Electron. Letters*, **4** (14), 298–299, 1968.
32. G. Kishi and T. Kida, 'Energy theory of sensitivity in *LCR* networks', *IEEE Trans.*, **CT-14** (4), 380–387, 1967.

33. W. E. Smith, 'Element sensitivity and energy storage of a passive impedance', *IEEE Trans.*, **CT-18** (3), 337–342, 1971.
34. M. L. Blostein, 'Some bounds on sensitivity in RLC networks', in *Proceedings of the 1st Allerton Conference on Circuits and Systems Theory*, University of Illinois, Urbana, USA; 1963, pp. 488–501.
35. J. D. Schoeffler, 'The synthesis of minimum sensitivity networks', *IEEE Trans.*, **CT-11**, 270–276, 1964.
36. L. T. Bruton, 'Sensitivity comparison of high-Q second-order active filter synthesis techniques', *IEEE Trans.*, **CAS-22** (1), 32–38, 1975.
37. R. W. Newcomb, 'The non-completeness of continuously equivalent networks', *IEEE Trans.*, **CT-13**, 207–208, 1966.
38. B. A. Shenoi, 'Optimum variability design and comparative evaluation of thin-film RC-active filters', *IEEE Trans.*, **CAS-21**, 263–267, 1974.
39. C. L. Semmelman, E. D. Walsh, and G. Daryanani, 'Linear circuits and statistical design', *Bell Systems Tech. J.*, **50** (4), 1149–1171, 1971.
40. J. M. Hammersley and D. C. Handscomb, *Monte Carlo Methods*, Methuen, London, 1964.

5. Active elements

A variety of active elements are used in the realization of inductorless filters. These include controlled sources, converters, and inverters. Although such functions can be implemented by direct integrated designs, this is not usually feasible unless large-scale production is envisaged. Instead, emphasis is laid mainly on their indirect realization by means of a special degenerate element: namely, the voltage operational amplifier. This widely-used practical component (whose direct applications are discussed in Chapter 7) is readily available in integrated form and possesses well-defined properties which simplify the realization of active building blocks and render multiple-amplifier methods economically feasible. Throughout this chapter, therefore, the utility of the operational amplifier in providing viable active elements is stressed. The range of active components is seen to be represented mathematically in terms of different simplifications of the basic transmission matrix; the ideal operational amplifier being the completely null case.

5.1 Pathological elements

To assist in achieving a generalized approach, it is helpful at this juncture to consider elements having terminal behaviour which cannot be described by any conventional matrix representations. On account of such abnormalities, these degenerate forms are usually called 'pathological' elements and are necessary to complete and unify the range of fundamental network elements.[1]

The basic set of active elements comprises:

(a) The *nullator:* a two-terminal element defined by $V = I = 0$ and simultaneously presenting an open-circuit and a short-circuit (symbolized in Fig. 5.1(a)). It is, in fact, the 'virtual earth' concept[2] inherent in operational amplifiers.

(b) The *norator:* a two-terminal element for which V and I are completely arbitrary (symbolized in Fig. 5.1(b)).

(c) The *nullor:* a two-port pathological network made up of a nullator at one port and a norator at the other (see Fig. 5.1(c)). It is described by

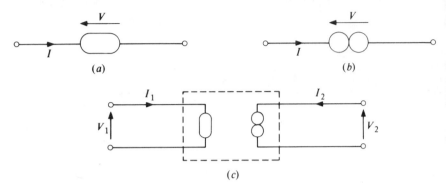

Fig. 5.1. Pathological elements. (*a*) Nullator symbol; (*b*) norator symbol; (*c*) nullor symbol.

the null transmission matrix

$$\begin{bmatrix} V_1 \\ I_1 \end{bmatrix} = \begin{bmatrix} 0 & 0 \\ 0 & 0 \end{bmatrix} \begin{bmatrix} V_2 \\ I_2 \end{bmatrix} \qquad (5.1)$$

A monolithic nullor with floating input and output ports has recently been described.[3] The eventual availability of such universal active elements affords the possibility of a floating output port as well as minimization of the number of active devices and precision passive elements in practical circuits.

The nullator and norator are self duals and have a role in the representation of physical devices such as the bipolar transistor and operational amplifier as well as in the generation and manipulation of equivalent circuits. In this manner, new and frequently useful circuit configurations can be obtained. Since all voltages and currents in physically realizable circuits must be defined the nullator and norator must always occur as a pair. This is demonstrated by the ideal common-base junction transistor equivalent of Fig. 5.2.

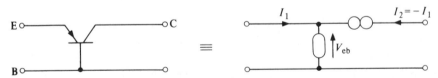

Fig. 5.2. Pathological equivalent of common-base junction transistor

When using pathological elements, alternative network realizations can be derived by converting and reconverting from the pathological form. Whenever a nullator and norator occur in series an open circuit results, while, if a nullator

Fig. 5.3. Simple pathological combinations. (*a*) Open circuit; (*b*) short circuit

ACTIVE ELEMENTS

and norator occur in parallel, a short circuit results (Fig. 5.3). These concepts are helpful in the sections which follow for the characterization of the variety of ideal active elements to be introduced.

5.2 Operational amplifiers

Although the voltage operational amplifier has received most attention, present design practice allows the engineer to realize other forms of operational amplifier such as the current, transconductance, and transimpedance types.

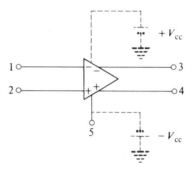

Fig. 5.4. General five-terminal amplifier

In its most general form, the operational amplifier may be represented[4] by the five-terminal device shown in Fig. 5.4. Numbers 1, 2, 3 and 4 represent the input and output noninverting signal terminals, respectively. The practical direct-voltage supplies $\pm V_{cc}$ are shown dotted, being the effective a.c. ground of terminal 5. Using this as the reference node and assuming the internal feedback terms to be zero, the definite admittance matrix may be written as

$$[Y] = \begin{bmatrix} Y_{11} & -Y_{12} & 0 & 0 \\ -Y_{21} & Y_{22} & 0 & 0 \\ -Y_{31} & -Y_{32} & Y_{33} & -Y_{34} \\ -Y_{41} & -Y_{42} & -Y_{43} & Y_{44} \end{bmatrix} \quad (5.2)$$

In the ideal case, it can be assumed that the differential input admittance $Y_{12} = Y_{21} \rightarrow 0$. Similarly, the differential output admittance $Y_{34} = Y_{43} \rightarrow 0$.

By defining the following differential transfer parameters:

$$A = \text{open-circuit voltage gain} = \left.\frac{V_4}{V_2 - V_1}\right|_{I_4=0} = \left.\frac{V_3}{V_1 - V_2}\right|_{I_3=0}$$

$$B = \text{short-circuit current gain} = \left.\frac{I_4}{I_1 - I_2}\right|_{V_4=0} = \left.\frac{I_3}{I_2 - I_1}\right|_{V_3=0}$$

$$C = \text{transfer admittance} = \frac{I_3}{V_2 - V_1}\bigg|_{V_3=0} = \frac{I_4}{V_1 - V_2}\bigg|_{V_4=0}$$

$$D = \text{transfer impedance} = \frac{V_4}{I_2 - I_1}\bigg|_{I_4=0} = \frac{V_3}{I_1 - I_2}\bigg|_{I_3=0}$$

it is possible to represent the various admittance matrices in terms of these parameters and the driving-point admittances as shown in Table 5.1. Since

Table 5.1 Admittance matrices for five-terminal amplifiers

Amplifier type		Admittance matrix			
		1	2	3	4
Voltage	1	0	0	0	0
	2	0	0	0	0
	3	$-AY_{33}$	AY_{33}	Y_{33}	0
	4	AY_{44}	$-AY_{44}$	0	Y_{44}
Current	1	Y_{11}	0	0	0
	2	0	Y_{22}	0	0
	3	$-BY_{11}$	BY_{22}	0	0
	4	BY_{11}	$-BY_{22}$	0	0
Transadmittance	1	0	0	0	0
	2	0	0	0	0
	3	$-C$	C	0	0
	4	C	$-C$	0	0
Transimpedance	1	Y_{11}	0	0	0
	2	0	Y_{22}	0	0
	3	$-DY_{11}Y_{33}$	$DY_{22}Y_{33}$	Y_{33}	0
	4	$DY_{11}Y_{44}$	$-DY_{22}Y_{44}$	0	Y_{44}

voltage-type operational amplifiers have had widest application in active filters, they will now be considered in more detail. Outside the present section, the term 'OA' should be understood to mean 'voltage operational amplifier'. Other types of amplifier receive further consideration in Sec. 5.3.

The most frequently used single-ended version of the operational amplifier is shown in Fig. 5.5(a), together with its ideal pathological equivalent in Fig. 5.5(b) and its simplified admittance circuit model in Fig. 5.5(c). It has been assumed that there is no transfer admittance between input and output and that the gain is independent of frequency (see Chapter 9 for effects of frequency dependence).

Analysis of Fig. 5.5(c) yields the following equations:

$$I_1 = Y_{11}V_1$$

$$I_4 = AY_{44}V_1 + Y_{44}V_4$$

ACTIVE ELEMENTS

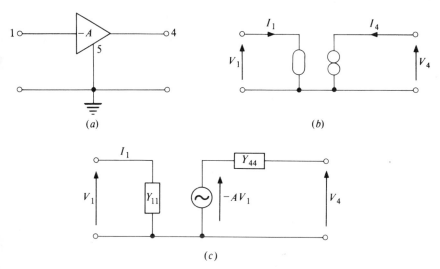

Fig. 5.5. Three-terminal phase-reversing operational amplifier. (*a*) Symbol; (*b*) ideal pathological equivalent; (*c*) circuit model

or, in matrix form,

$$[I] = \begin{bmatrix} Y_{11} & 0 \\ AY_{44} & Y_{44} \end{bmatrix} [V] \qquad (5.3)$$

The above matrix can be used to represent a practical phase-reversing amplifier and is particularly suited for use in computer-based analysis when A is a real number. Table 5.2 illustrates the network symbols, models, and matrices for a number of amplifier terminal arrangements. The numbering is consistent with that in Table 5.1 and the models can be seen to include the input common-mode and differential input admittances hitherto omitted in the general amplifier comparisons. Commercial devices are available to realize all entries in Table 5.2, but the most common are four-terminal amplifiers with differential-input and single-ended output terminals. Differential-output five-terminal types are rarely used in filter design because of the lower open-loop gain and matching problems of the paraphase output.

5.3 Controlled sources

Controlled sources are two-port unidirectional devices having an output (or controlled) variable dependent upon an input (or controlling) variable and independent of any other variables in the network. The four basic types of controlled source are shown in idealized form in Table 5.3, together with their matrix description, pathological equivalent, and a practical example of each

Table 5.2 Models for voltage operational amplifiers

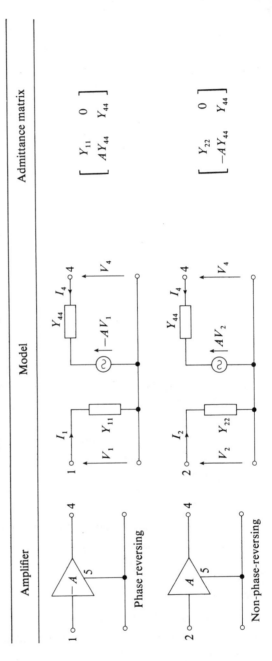

ACTIVE ELEMENTS

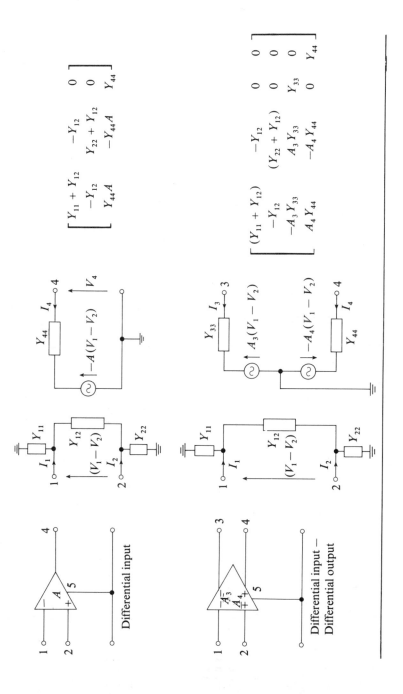

ACTIVE FILTERS FOR COMMUNICATIONS AND INSTRUMENTATION

Table 5.3 Ideal controlled sources

Controlled source type	Ideal model	Matrix description	Pathological equivalent	Idealized practical example
Voltage-controlled voltage source (VCVS)	KV_1	$\begin{bmatrix} I_1 \\ V_2 \end{bmatrix} = \begin{bmatrix} 0 & 0 \\ K & 0 \end{bmatrix} \begin{bmatrix} V_1 \\ I_2 \end{bmatrix}$ $[G]$	$\left(1 + \dfrac{R_2}{R_1}\right) = K$	FET
Voltage-controlled current source (VCCS)	GV_1	$\begin{bmatrix} I_1 \\ I_2 \end{bmatrix} = \begin{bmatrix} 0 & 0 \\ G & 0 \end{bmatrix} \begin{bmatrix} V_1 \\ V_2 \end{bmatrix}$ $[Y]$	$R_1 = 1/G$	Pentode
Current-controlled current source (CCCS)	KI_1	$\begin{bmatrix} V_1 \\ I_2 \end{bmatrix} = \begin{bmatrix} 0 & 0 \\ -K & 0 \end{bmatrix} \begin{bmatrix} I_1 \\ V_2 \end{bmatrix}$ $[h]$	$\left(1 + \dfrac{R_2}{R_1}\right) = -K$	Bipolar transistor
Current-controlled voltage source (CCVS)	RI_1	$\begin{bmatrix} V_1 \\ V_2 \end{bmatrix} = \begin{bmatrix} 0 & 0 \\ R & 0 \end{bmatrix} \begin{bmatrix} I_1 \\ I_2 \end{bmatrix}$ $[Z]$	$R_2 = -R$	d.c. rotary generator

ACTIVE ELEMENTS

source; they have finite-real control constants and may be derived from the five-terminal active device discussed in Sec. 5.2. Hence:

(a) the voltage-controlled voltage source (VCVS) is an element with output voltage controlled by input voltage and may be derived from the voltage amplifier;

(b) the voltage-controlled current source (VCCS) is an element with output current controlled by input voltage and may be derived from the transadmittance amplifier;

(c) the current-controlled current source (CCCS) is an element with output current controlled by input current and may be derived from the current amplifier;

(d) the current-controlled voltage source (CCVS) is an element with output voltage controlled by input current and may be derived from the transimpedance amplifier.

Since the above controlled sources have zero input power requirement (either input voltage or input current is zero) and can deliver unlimited output power, they exhibit infinite power gain. Incidentally, it is strictly necessary only to define two devices[5] since the others may be obtained by cascading. This is illustrated in the following example.

Example 5.1 Realize each of the following controlled sources as a cascade of two other controlled sources:

(i) VCVS (ii) CCCS

SOLUTIONS (i) The required source suggests a cascade of a VCCS and a CCVS, as shown in Fig. 5.6(a). Since $V = 0$ for a nullator, the input voltage appears across R_1, and the nodal equation is:

$$\frac{V_1}{R_1} + I' = 0$$

The current I' must flow through R_2 since no part of it can flow through the second nullator. In addition, since $V' = 0$, then I' is related to the output voltage V_2 by the equation

$$I' = -\frac{V_2}{R_2}$$

Thus, from the two expressions for I',

$$V_2 = \left(\frac{R_2}{R_1}\right) V_1$$

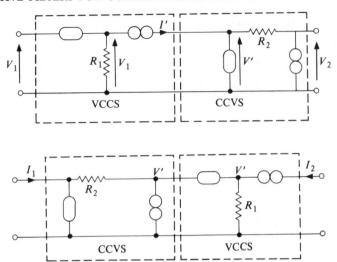

Fig. 5.6. Controlled sources resulting from cascade of pathological networks. (*a*) VCVS; (*b*) CCCS

which clearly represents a VCVS, though with a different gain from the model in Table 5.3.

(*ii*) In this case the required source suggests a cascade of a CCVS and a VCCS, as shown in Fig. 5.6(*b*). Due to the presence of the input nullator, $V_1 = 0$ and $V' = -I_1 R_2$.

The second nullator establishes V' across R_1 and, since I_2 must flow through R_1,

$$V' = I_2 R_1$$

The two expressions for V' may be equated to yield

$$I_2 = -\left(\frac{R_2}{R_1}\right) I_1$$

which clearly represents a CCCS though, once again, with a different gain from the model in Table 5.3.

Practical (non-ideal) controlled sources are now discussed in more detail, analysis being based on the admittance matrix since this lends itself most readily to computer-assisted studies.

5.3.1 Voltage-controlled voltage source (VCVS)

This device is easily realized using an operational amplifier and forms the basis for a variety of active-filter structures (see Chapter 6). The controlled source (or voltage amplifier) may have positive or negative gain determined, to a first

ACTIVE ELEMENTS

approximation, by the feedback elements. The most commonly used VCVS circuits are shown in Table 5.4, together with their corresponding admittance matrices.

Considering the basic feedback inverting network, the expression for the voltage gain may be obtained from the admittance matrix as

$$\frac{V_4}{V_s} = \frac{-y_{4s}}{y_{44}} = \frac{-Y_a(Y_{44} - Y_b/A)}{Y_b Y_{44} + (1/A)\{(Y_{44} + Y_b)(Y_{11} + Y_a) + Y_b Y_{44}\}} \quad (5.4)$$

If Y_{11} and $1/A \to 0$, the expression reduces to the more familiar

$$\frac{V_4}{V_s} = K = -\left(\frac{Y_a}{Y_b}\right)$$

Real, finite values of gain may thus be obtained as the ratio of two resistors.
The corresponding input admittance is

$$Y_{in} = Y_{ss} - \frac{Y_{s4} Y_{4s}}{Y_{44}} \quad (5.5)$$

Using the matrix elements of Table 5.4, the reader may wish to verify that, as A and $Y_{44} \to \infty$ while $Y_{11} \to 0$, then $Y_{in} \to Y_a$, which is a well-known result.[2] Care must be exercised when using inverting controlled sources so as to ensure that Y_a is small enough to avoid loading effects on the preceding stage or source.

The output admittance Y_{out} is (from the first matrix in Table 5.4)

$$Y_{out} = Y_{44} + Y_b + \frac{Y_b(AY_{44} - Y_b)}{Y_{11} + Y_a + Y_b}$$

For $Y_{11} \to 0$ and $Y_{44} \gg Y_b$, this reduces to

$$Y_{out} = Y_{44}\left\{1 + \frac{AY_b}{Y_a + Y_b}\right\} = \frac{Y_{44} A}{(1 + Y_a/Y_b)} \quad (5.6)$$

if

$$\frac{AY_b}{(Y_a + Y_b)} \gg 1.$$

Equation (5.6) shows that the output admittance is greatly increased by the addition of negative feedback. For instance, the popular μA741-type operational amplifier with a typical output resistance of 75 Ω ($1/Y_{44}$) and an open-loop gain (A) of 10^3 at 1 kHz will have an effective output resistance of $(1 + Y_a/Y_b) \times 75 \times 10^{-3}$ Ω. Thus, for a controlled gain ($-Y_a/Y_b$) of -9, the output resistance will be a mere 0.75 Ω! The effects of this parameter can clearly be ignored.

Table 5.4 Practical realizations and parameters for VCVS's

Source type	Source representation	Two-port admittance matrix	Voltage gain ($A = \infty$)	Input admittance (Y_{in})	Output admittance (Y_{out})
Inverting		$\begin{bmatrix} \left(Y_a - \dfrac{Y_a^2}{Y_{11}+Y_a+Y_b}\right) & \dfrac{-Y_a Y_b}{Y_{11}+Y_a+Y_b} \\ \dfrac{Y_a(AY_{44}-Y_b)}{Y_{11}+Y_a+Y_b} & \left\{(Y_{44}+Y_b)+\dfrac{Y_b(AY_{44}-Y_b)}{Y_{11}+Y_a+Y_b}\right\} \end{bmatrix}$	$-\dfrac{Y_a}{Y_b}$	Y_a	$\dfrac{AY_{44}}{(1+Y_a/Y_b)}$
Non-inverting		$\begin{bmatrix} (Y_{22}+Y_{12}) - \dfrac{-Y_{12}^2}{Y_{11}+Y_{12}+Y_a+Y_b} & \dfrac{-Y_{12}Y_b}{Y_{11}+Y_{12}+Y_a+Y_t} \\ -AY_{44}+\dfrac{Y_{12}(AY_{44}-Y_b)}{Y_{11}+Y_{12}+Y_a+Y_b} & (Y_{44}+Y_b)+\dfrac{Y_b(AY_{44}-Y_b)}{Y_{11}+Y_{12}+Y_a+Y_b} \end{bmatrix}$	$1+\dfrac{Y_a}{Y_b}$	$\dfrac{Y_{12}(1+Y_a/Y_b)}{A}$	$\dfrac{AY_{44}}{(1+Y_a/Y_b)}$
Voltage follower		$\begin{bmatrix} Y_{12} & -Y_{12} \\ -(AY_{44}+Y_{12}) & Y_{44}(1+A)+Y_{12} \end{bmatrix}$	1	0	∞

The above results have been obtained by use of the nodal admittance matrix and the reader may care to verify them using conventional mesh-nodal analysis.[2] Results for all three VCVS's are summarized in Table 5.4. It will be seen that resistor ratios alone are responsible for establishing the gain in the first two cases. This fact is of particular importance in integrated-circuit fabrication since tolerances on absolute values of resistance are much greater than ratio tolerances. It should be noted that the non-inverting arrangements require a differential-input device since they use the common-mode effect in the feedback arrangement.

The voltage follower is simply a limiting case of the non-inverting amplifier. With zero feedback resistance, the output 'follows' the input (the emitter and source-follower circuits are non-ideal versions of the voltage follower) and the gain is

$$K = \frac{A}{1 + A} \tag{5.7}$$

which simplifies to unity when $A = \infty$. Non-inverting circuits with $Z_{in} \to \infty$ and $Z_{out} \to 0$ provide excellent impedance buffering, a useful practical facility. This is particularly true of the follower since it requires no external resistors.

5.3.2 Voltage-controlled current source (VCCS)

This type of controlled source (or transconductance amplifier) may be realized[2] by the operational-amplifier arrangement shown in Fig. 5.7(a). Circuit analysis

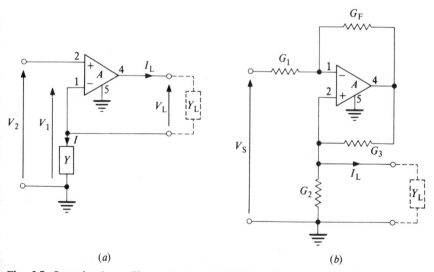

Fig. 5.7. Operational amplifier realizations of VCCS. (a) Floating output terminals; (b) grounded output terminals

gives

$$V_1 = I/Y \tag{5.8a}$$

$$V_L + V_1 = A(V_2 - V_1) \tag{5.8b}$$

$$I_L = -I_4 = I \tag{5.8c}$$

which, as $A \to \infty$, yield

$$I_L = YV_2 \tag{5.9}$$

describing a voltage-controlled current source, albeit with a floating output. The VCCS can also be realized with a grounded load, see Fig. 5.7(b), for which

$$I_L = G_2 V_S \tag{5.10}$$

under the condition that $G_1 G_3 = G_2 G_F$. Source terminal impedances will usually require high resistance values and consequently an FET-input operational amplifier (see Table 9.1). A further input buffer amplifier may also be necessary for integrated-circuit realization.

Transistor implementations are, however, generally more convenient for this type of controlled source and Fig. 5.8(a) shows a relatively simple transconductance amplifier ideally suited to fabrication in integrated form. The circuit comprises a composite transistor pair and elementary analysis yields I_2 as

$$I_2 = \alpha_2 I_0 \simeq \frac{\alpha_2 (V_{cc} - V_1)}{R}$$

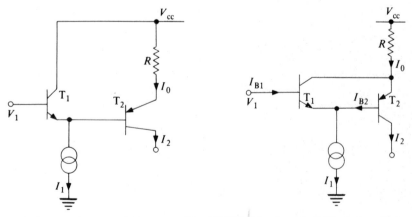

Fig. 5.8. Integrated transconductance amplifiers (VCCS), (a) α-dependent; (b) α-independent

ACTIVE ELEMENTS

if it is assumed that $V_{BE1} = V_{BE2}$. For a.c. signals, the device functions as a VCCS with transconductance G, given by

$$G = \frac{\partial I_2}{\partial V_1} = \frac{-\alpha_2}{R} \tag{5.11}$$

which has a 3 dB bandwidth as high as the alpha cut-off frequency (f_α) of the pnp transistor (typically in megahertz). A substantial improvement in bandwidth may be obtained by circulating a fraction of I_0 through the npn transistor, as shown in Fig. 5.8(b). Current continuity through the circuit requires that

$$I_{B1} + I_0 = I_2 + I_1 \tag{5.12}$$

which, when the current gain of T_1 is high ($I_{B1} \to 0$), reduces to

$$I_2 = I_0 - I_1 = \left(\frac{V_{cc} - V_1}{R}\right) - I_1$$

and, consequently,

$$G = \frac{\partial I_2}{\partial V_1} = \frac{-1}{R} \tag{5.13}$$

Thus, to a first-order, the transconductance is independent of α_2 and the 3 dB bandwidth of this positive-feedback arrangement can occur[6] as high as $3f_\alpha$. A more sophisticated version of the transconductance amplifier is produced in integrated form and will be discussed in Sec. 5.3.5.

5.3.3 Current-controlled current source (CCCS)

Bipolar transistors are essentially current-controlled current sources, though their performance falls short of the ideal. For example, the common-base arrangement cannot provide a current gain in excess of unity but its good isolation properties (low input impedance and high output impedance) can prove useful in the first stage of a current amplifier.

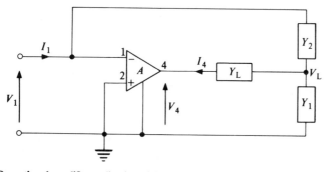

Fig. 5.9. Operational amplifier realization of CCCS

A CCCS realization using an OA is shown in Fig. 5.9, the nodal equations being

$$V_4 = -AV_1 \qquad (5.14a)$$
$$I_1 = (V_1 - V_L)Y_2 \qquad (5.14b)$$
$$I_4 = (V_L - V_4)Y_L \qquad (5.14c)$$
$$I_1 - I_4 = V_L Y_1 \qquad (5.14d)$$

which reduce to

$$I_1\left\{1 + \frac{Y_2}{Y_1} + \frac{Y_2^2}{AY_1Y_L}\right\} = I_4 Y_2\left\{\frac{1}{Y_1} + \frac{1}{AY_L}\left(1 + \frac{Y_2}{Y_1}\right)\right\}$$

As $A \to \infty$,

$$I_4 = \left(1 + \frac{Y_1}{Y_2}\right)I_1 \qquad (5.15)$$

and is clearly of the required form. The practical disadvantage of this circuit is the floating load, Y_L. For a grounded load, multiple-amplifier simulations or direct transistor implementations of the simple source or its current-conveyor generalization are necessary.

An alternative circuit designed to take advantage of monolithic fabrication is shown in Fig. 5.10. It is well known that two transistors operated at the same V_{BE} have emitter currents related according to their respective areas. The current in the diode-connected transistor is controlled by V_{cc} and R so that

$$I_1 = \frac{(V_{cc} - V_{BE1})}{R} \approx I_{E1}$$

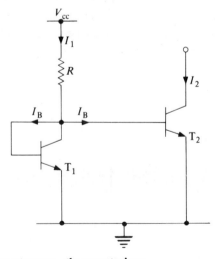

Fig. 5.10. Integrated current source—the current mirror

Since $V_{BE1} = V_{BE2}$, the current I_2 is controlled as

$$I_2 = \left(\frac{V_{cc} - V_{BE1}}{R}\right) \times \left(\begin{array}{c}\text{Ratio of}\\ \text{emitter areas}\end{array}\right) \qquad (5.16)$$

neglecting the effects of base current. For equal emitter cross-sections, $I_2 = I_1$ and the circuit is known[6] as a 'current mirror'. With appropriate bias, this device can be used as a CCCS since the input current variations will control those at the output.

5.3.4 Current-controlled voltage source (CCVS)

This is also referred to as a current-to-voltage converter or transducer and a transimpedance amplifier since it has a forward transfer impedance and no

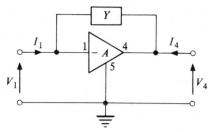

Fig. 5.11. Current-controlled voltage source (CCVS)

reverse transmission. It may readily be implemented using a single operational amplifier as in Fig. 5.11 with matrix equation

$$\begin{bmatrix} I_1 \\ I_4 \end{bmatrix} = \begin{bmatrix} Y + Y_{11} & -Y \\ AY_{44} - Y & Y + Y_{44} \end{bmatrix} \begin{bmatrix} V_1 \\ V_4 \end{bmatrix} \qquad (5.17)$$

The required transfer impedance is then given by

$$\left.\frac{V_4}{I_1}\right|_{V_1=0} = -\left(\frac{1}{Y}\right) \qquad (5.18)$$

The input admittance may be determined with the aid of Eq. (5.5) as

$$Y_{in} = \frac{(Y_{11} + Y)(Y + Y_{44}) + Y(AY_{44} - Y)}{(Y + Y_{44})}$$

If $Y_{11} \to 0$ and $Y_{44} \to \infty$, this reduces to the familiar expression

$$Y_{in} = Y(1 + A) \qquad (5.19)$$

The CCVS has not received much attention in the synthesis of active filters and does not, in this text, receive further discussion except for its application in frequency-emphasizing networks (Sec. 6.3.2).

ACTIVE FILTERS FOR COMMUNICATIONS AND INSTRUMENTATION

5.3.5 Multi-terminal controlled sources

The definition of a controlled source given earlier (Sec. 5.3) can be extended to include devices with more than one controlling variable. An example, in the

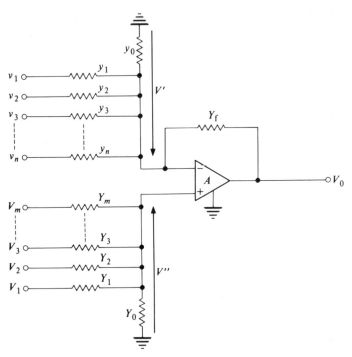

Fig. 5.12. Multiple-input VCVS

form of a VCVS, is illustrated in Fig. 5.12. Analysis may be performed by superposition. Thus, for the inverting inputs only,

$$(v_1 - V')y_1 + (v_2 - V')y_2 + \cdots + (v_n - V')y_n - V'y_0 = (V' - V_0)Y_f$$

and

$$V' = -V_0/A$$

This leads quite simply to

$$V_0 = -\sum_{i=1}^{n} v_i y_i \left[\frac{A}{Y_f(1+A) + y} \right] \quad (5.20)$$

where

$$y = \sum_{i=0}^{n} y_i \quad (5.21)$$

Now, for the non-inverting inputs,

$$(V_1 - V'')Y_1 + (V_2 - V'')Y_2 + \cdots + (V_m - V'')Y_m = V''Y_0$$

and

$$V_0 = (V'' - V')A = \left(V'' - \frac{V_0 Y_f}{y + Y_f}\right)A$$

This may easily be simplified and added to the result in Eq. (5.20) to yield a total output given by

$$V_0 = -\sum_{i=1}^{n} v_i A \left[\frac{y_i}{Y_f(1 + A) + y}\right] + \sum_{i=1}^{m} \frac{AV_i Y_i}{Y}\left(\frac{y + Y_f}{Y_f(1 + A) + y}\right) \quad (5.22)$$

where

$$Y = \sum_{i=0}^{m} Y_i \quad (5.23)$$

Although this configuration has been most frequently employed as an inverting summer ($V_i = 0$), it is a useful device in analogue computing (state-variable) synthesis (Chapters 8 and 9) for providing both positive and negative gain.

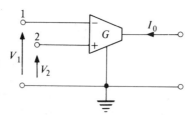

Fig. 5.13. Symbol for differential voltage-controlled current source (DVCCS)

A differential voltage-controlled current source (DVCCS) is shown in symbolic form in Fig. 5.13. In conjunction with a capacitor, this device can yield all components necessary for linear-circuit construction.[7] Its performance is described by the following terminal equation:

$$I_0 = G(V_2 - V_1) \quad (5.24)$$

with $G \geqslant 0$.

An integrated version of the DVCCS is available commercially (RCA 3080) and is referred to as an operational transconductance amplifier. The setting of the bias current I_B determines[8] the transconductance as

$$G = (19.2) I_B \quad \text{S} \quad (5.25)$$

where $0.1 \, \mu A < I_B < 1$ mA at 25°C. It is worth noting that the basic building block in the device is a current mirror (as described in Sec. 5.3.3), the i.c. chip incorporating 94 transistors and not a single resistor.

Combined with an output buffer, the transconductance amplifier could become a useful practical component in active filters. The hybrid device, known as the DVCCS/DVCVS (a terminology which persists despite the single-ended nature of the VCVS), is shown in symbolic form in Fig. 5.14, its performance being defined by the following set of equations:

$$I_0 = G(V_2 - V_1) \tag{5.26a}$$

$$V_0 = KV_g = \frac{KG}{G_1} V(V_2 - V_1) \tag{5.26b}$$

assuming the parallel combination of DVCCS output conductance and DVCVS input conductance, $G_1 \to 0$.

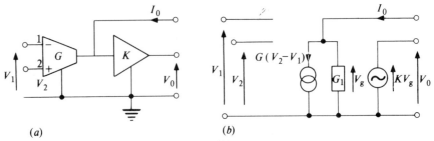

Fig. 5.14. DVCCS/DVCVS. (a) circuit symbol; (b) device model

This is an example of a controlled source with more than one controlled and controlling variable. It will be further discussed in Chapter 6 as a synthesis element. In integrated form, it can be produced on a single chip and unity-buffered versions ($\alpha = 1$) may be adapted from certain commercially available amplifiers (e.g., RCA 3140) by utilizing the strobe terminal as the current-source output.[7]

5.4 Immittance converters and inverters

An ideal impedance converter may be defined as a two-port device which, when terminated at one port by an impedance Z_L, presents an input impedance proportional to Z_L at the other port. The essential requirements for such devices may be established with the aid of Fig. 5.15. The composite $[a]$ matrix is given by

$$[a'] = \begin{bmatrix} a'_{11} & a'_{12} \\ a'_{21} & a'_{22} \end{bmatrix} = \begin{bmatrix} a_{11} & a_{12} \\ a_{21} & a_{22} \end{bmatrix} \begin{bmatrix} 1 & 0 \\ 1/Z_L & 1 \end{bmatrix} = \begin{bmatrix} a_{11} + a_{12}/Z_L & a_{12} \\ a_{21} + a_{22}/Z_L & a_{22} \end{bmatrix}$$

ACTIVE ELEMENTS

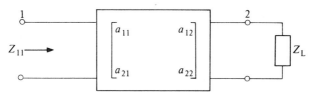

Fig. 5.15. Network for establishing properties of impedance converters

The input impedance may be determined as

$$Z_{11} = \frac{a'_{11}}{a'_{21}} = \frac{a_{11}Z_L + a_{12}}{a_{21}Z_L + a_{22}} \qquad (5.27)$$

Similarly, if port 1 is terminated by Z_L,

$$Z_{22} = \frac{Z_L a_{22} + a_{12}}{Z_L a_{21} + a_{11}} \qquad (5.28)$$

The two-port will act as an impedance converter if $a_{12} = a_{21} = 0$, for then

$$Z_{in_1} = Z_{11} = \frac{a_{11}}{a_{22}} Z_L = k(s) Z_L \qquad (5.29a)$$

and

$$Z_{in_2} = Z_{22} = \frac{a_{22}}{a_{11}} Z_L = \frac{Z_L}{k(s)} \qquad (5.29b)$$

where $k(s)$ is known as the conversion factor.

In terms of admittances, Eq. (5.29) may be written as

$$Y_{in_1} = \frac{a_{22}}{a_{11}} Y_L = \frac{Y_L}{k(s)} \qquad (5.30a)$$

and

$$Y_{in_2} = \frac{a_{11}}{a_{22}} Y_L = k(s) Y_L \qquad (5.30b)$$

Negative impedance converters If $k(s)$ is real and negative, the device is referred to as a negative impedance converter (NIC). With reference to the matrix conversion table (Appendix 5), $a_{12} = a_{21} = 0$ implies that

$$h_{11} = h_{22} = 0 \quad \text{and if} \quad k(s) = -k, \quad h_{21} h_{12} = k$$

Two basic forms of NIC result, depending on whether the voltage or current suffers a phase reversal. Hence:

(a) *Current (CNIC)*, for which $h_{12} = 1$, $h_{21} = k$, so that the terminal equations are

$$V_1 = V_2, \qquad I_2 = k I_1$$

(b) *Voltage* (*VNIC*), for which $h_{12} = -k$, $h_{21} = -1$, so that the terminal equations are

$$V_1 = -kV_2, \qquad I_2 = -I_1$$

The NIC almost invariably has a conversion factor of unity.

Positive impedance converters If $k(s)$ is real and positive, the device is referred to as a positive impedance converter (PIC). An important example of this converter is the ideal transformer (IT) for which $Z_{11} = n^2 Z_L$, where n is the turns ratio. Note that the transformer is not necessarily formed from a pair (or set, if more than two coils) of magnetically-coupled coils and only the non-availability of high-power semiconductor devices restricts wider application of active transformers.

Since (using conventional transformer terminology) the coils may be wound in like or unlike fashion, the terminal voltages may be in or out-of-phase. This is reflected in the transmission matrix shown in Table 5.5.

Table 5.5 [*a*] **and** [*h*] **matrices for ideal converters**

Device	[*a*]	[*h*]
VNIC	$\begin{bmatrix} -k & 0 \\ 0 & 1 \end{bmatrix}$	$\begin{bmatrix} 0 & -k \\ -1 & 0 \end{bmatrix}$
CNIC	$\begin{bmatrix} 1 & 0 \\ 0 & -1/k \end{bmatrix}$	$\begin{bmatrix} 0 & 1 \\ k & 0 \end{bmatrix}$
IT	$\begin{bmatrix} \pm n & 0 \\ 0 & \mp 1/n \end{bmatrix}$	$\begin{bmatrix} 0 & \pm n \\ \mp n & 0 \end{bmatrix}$
VGIC	$\begin{bmatrix} \pm k(s) & 0 \\ 0 & 1 \end{bmatrix}$	$\begin{bmatrix} 0 & \pm k(s) \\ -1 & 0 \end{bmatrix}$
CGIC	$\begin{bmatrix} 1 & 0 \\ 0 & \pm 1/k(s) \end{bmatrix}$	$\begin{bmatrix} 0 & 1 \\ \mp k(s) & 0 \end{bmatrix}$

The ideal transformer is a lossless device since, with reference to Fig. 5.16,

$$V_1 I_1 + V_2 I_2 = V_1 I_1 + \frac{V_1}{n}(-nI_1) = 0$$

However, if $k = n_1 n_2$ and the device is satisfied by the equations

$$V_1 = \pm n_1 V_2, \qquad I_2 = \pm n_2 I_1$$

ACTIVE ELEMENTS

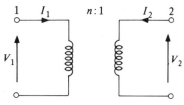

Fig. 5.16. Ideal transformer (IT)

then the losses can be negative if $n_2 > n_1$ and the device becomes an active transformer.

Generalized impedance converter If $k(s)$ (Eq. (5.29)) is complex, the device is referred to as a generalized impedance converter (GIC). As in the case of the NIC, there are two basic forms:

(a) *Current (CGIC)*, for which

$$V_1 = V_2, \qquad I_1 = \mp \left(\frac{1}{k(s)}\right) I_2$$

(b) *Voltage (VGIC)*, for which

$$V_1 = \pm k(s) V_2, \qquad I_1 = -I_2$$

The GIC will usually have a conversion factor of $k_1 s$ or $k_1 s^2$, though other factors would be legitimate. It should be noted that conversion factors of $1/k_1 s$ and $1/k_1 s^2$ may be obtained by using the device in the reverse direction. For this reason, both the conversion factor and port 1 are indicated, the latter by a dot, in the circuit symbol.[9] Thus, a GIC has a conversion factor $k_1 s^2$ when port 1 is the input port, and a conversion factor of $1/k_1 s^2$ when port 2 is the input port (Fig. 5.17(b)). Since the NIC conversion factor is invariably designed as unity, it is not customary to use the port 1 identification in its symbol. Converter symbols are shown in Fig. 5.17 while [a] and [h] matrix representations are included in Table 5.5.

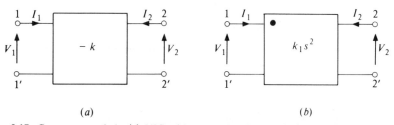

Fig. 5.17. Converter symbols. (a) NIC with conversion factor $-k$; (b) GIC with conversion factor $k_1 s^2$

The ideal impedance inverter may be defined as a two-port which, when terminated at one port by an impedance Z_L, presents an input impedance at the other port which is inversely proportional to Z_L. In terms of the transmission parameters (Eqs. (5.27) and (5.28)) the basic requirements of the two-port are:

$$a_{22} = a_{11} = 0$$

so that

$$Z_{11} = Z_{22} = \frac{a_{12}}{a_{21}} \left(\frac{1}{Z_L} \right) = \frac{G(s)}{Z_L} \tag{5.31}$$

Thus, an ideal impedance inverter has the same proportionality factor in each direction.

Negative impedance inverter When $G(s)$ is a negative real number the device is known as a negative impedance inverter (NII) which may ideally be defined by the following transmission matrix:

$$[a] = \begin{bmatrix} 0 & \pm 1/G \\ \mp G & 0 \end{bmatrix}$$

The device is represented by the nullator–norator arrangement shown in Fig. 5.18. The input current I_1 is forced to flow through R_1, developing a voltage such that

$$I_1 R_1 = -V_2$$

Similarly, I_2 flows through R_2, developing a voltage such that

$$V_1 = -I_2 R_2$$

and the transmission matrix may be written as

$$[a] = \begin{bmatrix} 0 & R_2 \\ -1/R_1 & 0 \end{bmatrix} \tag{5.32}$$

This is clearly of the required form and represents an ideal NII when $R_1 = R_2$. The arrangement will be used in the realization of gyrators in Sec. 5.4.2.

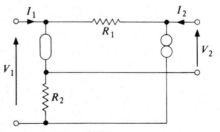

Fig. 5.18. Nullator-norator representation of NII

ACTIVE ELEMENTS

Positive impedance inverter When $G(s)$ is a positive real number, the inverter is called a positive impedance inverter (PII) or, more frequently, a gyrator. In the ideal case, it is described by

$$[a] = \begin{bmatrix} 0 & \pm 1/G \\ \pm G & 0 \end{bmatrix}$$

where G is referred to as the gyration conductance and G^2 is the gyration factor.

The circuit symbol for the ideal gyrator, together with its $[a]$, $[y]$, and $[z]$ matrices, is shown in Table 5.6. Notice that the immittance matrices are antisymmetric and hence the name 'gyrator' by analogy with the antisymmetric displacement property of a gyroscope spinning about a vertical axis. The significance of this concept was not exploited for many years after Tellegen's[10] enunciation of the theory. The ideal gyrator is a passive lossless device since

$$V_1 I_1 + V_2 I_2 = \left(\frac{1}{G} - \frac{1}{G}\right) I_1 I_2 = 0$$

However, though fundamentally reciprocal, practical realizations of the device inevitably require the use of non-reciprocal or active elements which render it non-ideal. In particular, $a_{12} \neq a_{21}$.

The following subsections (5.4.1 to 5.4.5) discuss practical realizations of impedance converters and inverters.

5.4.1 Negative impedance converters

The role of the NIC in the design of active filters[11] is more academic and historical than practical. It was particularly popular during the initial phase of intense activity in the development of active synthesis techniques. Its subsequent demise may be attributed to

(a) the high sensitivity to passive elements of the resulting configurations;
(b) its relative inconvenience for fabrication as a monolithic chip;
(c) the necessity of synthesizing from inherently sensitive polynomial difference decompositions (though this can be mitigated by optimization techniques).

However, occasional applications still occur (e.g., loss compensation in transmission lines and gyrator realization) while development of a cheap monolithic NIC may have advantages at high frequencies. Although no NIC synthesis techniques are discussed in this text, an example of a practical converter is considered.

Table 5.6 Circuit symbol and matrices for an ideal gyrator

Circuit symbol	Matrix description		
	$[a]$	$[y]$	$[z]$
(gyrator symbol)	$\begin{bmatrix} V_1 \\ I_1 \end{bmatrix} = \begin{bmatrix} 0 & \pm 1/G \\ \pm G & 0 \end{bmatrix} \begin{bmatrix} V_2 \\ -I_2 \end{bmatrix}$	$\begin{bmatrix} I_1 \\ I_2 \end{bmatrix} = \begin{bmatrix} 0 & \pm G \\ \mp G & 0 \end{bmatrix} \begin{bmatrix} V_1 \\ V_2 \end{bmatrix}$	$\begin{bmatrix} V_1 \\ V_2 \end{bmatrix} = \begin{bmatrix} 0 & \mp 1/G \\ \pm 1/G & 0 \end{bmatrix} \begin{bmatrix} I_1 \\ I_2 \end{bmatrix}$

ACTIVE ELEMENTS

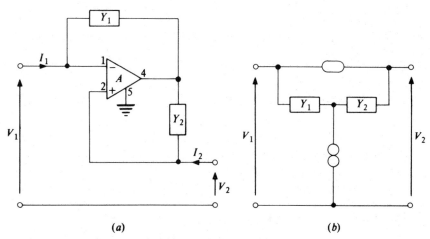

Fig. 5.19. CNIC. (a) Practical arrangement; (b) pathological equivalent

The nodal admittance matrix for the CNIC shown in Fig. 5.19(a) may be constructed with the aid of the differential-input amplifier matrix (Table 5.2) as

$$[Y] = \begin{array}{c} \\ 1 \\ 2 \\ 4 \end{array} \begin{array}{|ccc|} \hline 1 & 2 & 4 \\ \hline Y_{11} + Y_{12} + Y_1 & -Y_{12} & -Y_1 \\ -Y_{12} & Y_{22} + Y_{12} + Y_2 & -Y_2 \\ AY_{44} - Y_1 & -AY_{44} - Y_2 & Y_{44} + Y_1 + Y_2 \\ \hline \end{array}$$

which may be reduced by pivotal condensation about node 4 to:

$$[Y] = \begin{array}{c} \\ 1 \\ \\ 2 \end{array} \begin{array}{|cc|} \hline 1 & 2 \\ \hline Y_{11} + Y_{12} + Y_1 + \dfrac{Y_1(AY_{44} - Y_1)}{Y_{44} + Y_1 + Y_2} & -Y_{12} - \dfrac{Y_1(Y_2 + AY_{44})}{Y_{44} + Y_1 + Y_2} \\ -Y_{12} + \dfrac{Y_2(AY_{44} - Y_1)}{Y_{44} + Y_1 + Y_2} & Y_{22} + Y_{12} + Y_2 - \dfrac{Y_2(Y_2 + AY_{44})}{Y_{44} + Y_1 + Y_2} \\ \hline \end{array}$$

As Y_{44} and $A \to \infty$, the matrix may be simplified to

$$[Y] = \begin{bmatrix} AY_1 & -AY_1 \\ AY_2 & -AY_2 \end{bmatrix}$$

This can be converted to the transmission matrix for comparison with the ideal result shown in Table 5.5 as

$$[a] = \begin{bmatrix} 1 & 0 \\ 0 & -Y_1/Y_2 \end{bmatrix}$$

representing a CNIC with a conversion factor of $-Y_2/Y_1$. The usual synthesis requirement of unity conversion factor necessitates equal resistors in this instance.

Replacing the ideal operational amplifier by its nullator–norator equivalent from Fig. 5.5(b) gives Fig. 5.19(b). The nullator ensures that $V_1 = V_2$ and, since it allows no current flow, the voltage drops across Y_1 and Y_2 are equal and opposite. Hence, the above transmission-matrix form emerges.

5.4.2 Positive impedance inverters (gyrators)

Consider a gyrator terminated at its output port by an impedance Z. The overall transmission matrix can be derived as

$$[a] = \begin{bmatrix} 0 & 1/G \\ G & 0 \end{bmatrix} \begin{bmatrix} 1 & 0 \\ 1/Z & 1 \end{bmatrix} = \begin{bmatrix} 1/GZ & 1/G \\ G & 0 \end{bmatrix}$$

and the input impedance is

$$Z_{11} = \frac{a_{11}}{a_{21}} = \frac{1}{G^2 Z} \tag{5.33}$$

which represents an inversion of the terminating impedance. When Z is a capacitor, as in Fig. 5.20, Z_{11} is an inductance of value (C/G^2) henries. In this way it is possible to simulate inductance values far in excess of conventional coil values, thereby simplifying the practical aspects of synthesis, especially at low frequencies. Even at high frequencies, a higher quality factor than for a coil inductance can be attained as the latter suffers from skin effects and core losses.

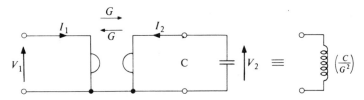

Fig. 5.20. Simulation of inductance by a gyrator

Pathological representation can be applied to any gyrator realization.[12] For example, in the parallel method, the gyrator $[Y]$ matrix is written as the sum of two matrices:

$$[Y] = \begin{bmatrix} 0 & G \\ -G & 0 \end{bmatrix} = \begin{bmatrix} 0 & G \\ 0 & 0 \end{bmatrix} + \begin{bmatrix} 0 & 0 \\ -G & 0 \end{bmatrix}$$

$$= [Y_1] + [Y_2]$$

ACTIVE ELEMENTS

Both $[Y_1]$ and $[Y_2]$ represent voltage-controlled current sources (VCCS) which, when connected in parallel, form a gyrator, as shown in Fig. 5.21. The resulting gyrator is non-ideal, the quality being limited due to the effect of phase shifts in the amplifiers. Negative feedback cannot easily be applied from a current output to a voltage input and compensation of the individual stages is necessary. Gyrators of this type have produced[13] medium-frequency Q factors in the range 500 to 1000. There are several other approaches: namely, series, negative resistance, series–parallel, and cascade. The latter technique is illustrated in the following example.

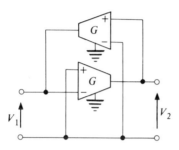

Fig. 5.21. Gyrator realization using two DVCCS's

Example 5.2 Show that a gyrator results from the cascade of a CNIC with a NII. Rearrange the pathological network so as to realize the gyrator with ideal differential-input OA's.

SOLUTION The transmission matrix resulting from the suggested cascade is given by

$$[a] = \begin{bmatrix} 1 & 0 \\ 0 & -1 \end{bmatrix} \begin{bmatrix} 0 & \pm R_2 \\ \mp 1/R_1 & 0 \end{bmatrix} = \begin{bmatrix} 0 & \pm R_2 \\ \pm 1/R_1 & 0 \end{bmatrix} \quad (5.34)$$

$$= [a]_{\text{CNIC}} \times [a]_{\text{NII}} = [a]_{\text{Gyrator}}$$

assuming the CNIC has unity conversion factor.

The pathological network representing Eq. (5.34) may be formed as the cascade of the pathological networks shown in Figs. 5.19(b) and 5.18. The arrangement is shown in Fig. 5.22(a). Rearrangement leads to the creation of two nullors (Fig. 5.22(b)) and the resulting amplifier realization is shown in Fig. 5.22(c). To facilitate understanding, the pathological elements are identified as NL_1, NL_2, NR_1, and NR_2 in the first two figures and the amplifier A_1 is formed from the nullor $NL_1:NR_2$ whilst amplifier A_2 is formed from the nullor $NL_2:NR_1$.

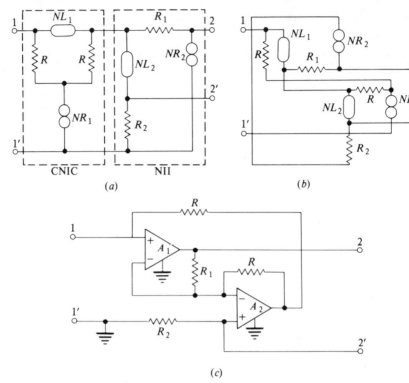

Fig. 5.22. Gyrator derived from pathological network. (a) Pathological-cascade network; (b) modified network forming nullors (NL_1, NR_2) and (NL_2, NR_1); (c) amplifier realization

The gyrator described in the above example was first proposed by Antoniou[13] and the matrix for the network can be shown to be

$$[Z] = K \begin{bmatrix} R(R_1[1+A_2] + R[1+A_1]) & -A_2R_2(R_1 + R[1+A_1]) \\ RA_1(R + R_1[1+A_2]) & R_2(R_1[1+A_2] + R[1+A_1]) \end{bmatrix} \quad (5.35)$$

where

$$K^{-1} = R_1(1+A_2) + R(1+A_1) + RA_1A_2$$

A gyrator of this type can produce[13] Q factors in the region of 1000 up to 2 kHz for inductor values between 0.1 and 1 H. Most high-quality gyrators require two OA's, though a single-amplifier realization[14] may occasionally suffice.

5.4.3 Ideal transformer

The IT with unity turns ratio ($n = 1$) can be realized as the cascade of two VNIC's, two CNIC's, two gyrators, or a VNIC and a CNIC. For example, using two CNIC's, the composite transmission matrix is

$$[a] = \begin{bmatrix} 1 & 0 \\ 0 & -1/k \end{bmatrix} \begin{bmatrix} 1 & 0 \\ 0 & -1/k \end{bmatrix} = \begin{bmatrix} 1 & 0 \\ 0 & 1/k^2 \end{bmatrix}$$

representing an IT if $k = 1$.

The restriction on the turns ratio can be relaxed if a generalized NIC (GNIC) is used,[14] for which the transmission matrix is

$$[a]_{GNIC} = \begin{bmatrix} k_1 & 0 \\ 0 & -1/k_1 \end{bmatrix}$$

Then the composite transmission matrix of the NIC cascade becomes

$$[a] = [a]_{GNIC} \times [a]_{CNIC} = \begin{bmatrix} k_1 & 0 \\ 0 & -1/k_1 \end{bmatrix} \begin{bmatrix} 1 & 0 \\ 0 & -1/k \end{bmatrix} = \begin{bmatrix} k_1 & 0 \\ 0 & 1/k_1 k \end{bmatrix}$$

and, for $k = 1$, the matrix represents an IT of turns ratio $n = k_1$.

The IT has practical limitations[15] in the form of frequency-band and impedance-range stability as well as the available output voltage and power-handling capacity. Nevertheless, it has obvious application when it is required to replace a conventional transformer by an equivalent device having no coupled coils. Apart from that, it has not attracted special attention, more interest having been centred on the device with a frequency-dependent turns ratio. This is known as the generalized impedance converter (GIC), to be discussed in the following subsection.

5.4.4 Generalized impedance converter

Since the GIC satisfies the same matrix description as the NIC (except that the conversion factor is complex instead of real), it is reasonable to suppose that some of their realized forms may be generally similar. The CNIC shown in Fig. 5.19(a) has a conversion factor given by

$$k = \frac{-Y_2}{Y_1} \qquad (5.36)$$

Hence, if $Y_1 = G_1$ and $Y_2 = sC_1$, the conversion factor is

$$k(s) = \frac{-sC_1}{G} \qquad (5.37)$$

which satisfies the requirements for a GIC. In practical applications, care must be taken in selecting the amplifier input terminals since one port of the GIC is open-circuit stable whilst the other is short-circuit stable.

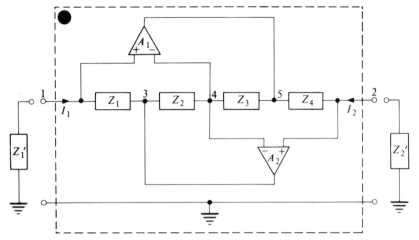

Fig. 5.23. Generalized impedance converter (GIC)

An extremely versatile realization of the GIC results from modifying the gyrator shown[13] in Fig. 5.22(c). If port 2 is terminated by an impedance and a new port 2 is created by removing R_2, the GIC shown in Fig. 5.23 results. For generality, all elements are shown as impedances. If it is assumed that the two amplifiers have infinite gain, the differential input voltage on each amplifier is zero and $V_1 = V_4 = V_2$. If the device is terminated at port 2 by an impedance Z'_2, the node equations may be written as

node 1 $V_1 - V_3 = Z_1 I_1$

node 4 $\left(\dfrac{V_3 - V_1}{Z_2}\right) + \left(\dfrac{V_5 - V_1}{Z_3}\right) = 0$

node 2 $-\dfrac{V_1}{Z'_2} + \left(\dfrac{V_5 - V_1}{Z_4}\right) = 0$

which reduce to

$$Z_{11} = \frac{V_1}{I_1} = \frac{Z_1 Z_3 Z'_2}{Z_2 Z_4}$$

Similarly, terminating port 1 with an impedance Z'_1 leads to

$$Z_{22} = \frac{V_2}{I_2} = \frac{Z_2 Z_4 Z'_1}{Z_1 Z_3}$$

The device thus acts as a CGIC with conversion factor

$$k(s) = \frac{Z_1 Z_3}{Z_2 Z_4} \qquad (5.38)$$

ACTIVE ELEMENTS

Depending upon the choice† of the impedances in Eq. (5.38), the device may be used to provide conversion factors of s^n, where $n = \pm 1, \pm 2$. If one internal impedance is capacitive and the other three are resistive, a resistive load at port 2 is converted to an inductive input impedance at port 1. With the same combination of internal elements, the conversion factor in the reverse direction is proportional to s^{-1} and a capacitor at port 1 produces a port 2 impedance given by $Z_{22} = (1/s^2 D)$, where D is a frequency invariant. At real frequencies, this represents a second-order 'capacitor' since

$$Z_{22}|_{s=j\omega} = \frac{-1}{\omega^2 D}$$

However, the term 'frequency-dependent negative resistance' (FDNR) seems to have achieved universal adoption.[17]

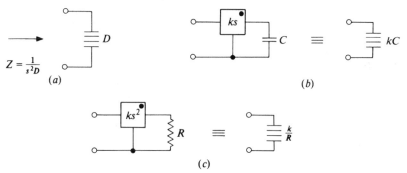

Fig. 5.24. FDNR symbol and realization. (a) Symbol for FDNR of value D; (b) realization of FDNR by capacitively terminated GIC; (c) realization of FDNR by resistively terminated GIC

D has units of (farad)² ohms and is represented by the symbol shown in Fig. 5.24(a). As indicated in Table 5.7, an FDNR may be created by a capacitively loaded GIC of conversion factor proportional to s^{-1} or by a resistively loaded GIC of conversion factor proportional to s^{-2}. The two methods of creating an FDNR are shown in Fig. 5.24(b) and (c).

Generalized impedance converters are available commercially. For instance, National Semiconductor Corporation manufactures a package (AF 120) containing a pair of operational amplifiers and four (matched) precision thin-film resistors. Capacitors have to be added as external components.

The GIC is a powerful addition to the building blocks available to filter designers and some of its applications are described in Chapter 8.

5.4.5 Multiport inverters and converters

Inverters and converters, being active two-ports, can be generalized to three or more ports in the same manner as controlled sources. Of particular interest is

† This was first suggested by Riordan[16] in considering the realization of gyrators.

ACTIVE FILTERS FOR COMMUNICATIONS AND INSTRUMENTATION

Table 5.7 Specific realizations using GIC

Element form				Load form		Input impedance		Element realized
Z_1	Z_2	Z_3	Z_4	Z_1'	Z_2'	Z_{11}	Z_{22}	
R_1	R_2	R_3	$\dfrac{1}{sC_4}$	$\dfrac{1}{sC_1'}$			$\dfrac{R_2}{s^2 C_4 C_1' R_1 R_3}$	FDNR
R_1	$\dfrac{1}{sC_2}$	R_3	$\dfrac{1}{sC_4}$	R_1'			$\dfrac{R_1'}{s^2 C_2 C_4 R_1 R_3}$	
R_1	R_2	R_3	$\dfrac{1}{sC_4}$		R_2'	$\dfrac{sC_4 R_1 R_3 R_2'}{R_2}$		Inductor
R_1	$\dfrac{1}{sC_2}$	R_3	R_4		R_2'	$\dfrac{sC_2 R_1 R_3 R_2'}{R_4}$		

the *n*-port generalized inverter (gyrator) which is a network with the description

$$\mathbf{i} = G\mathbf{v}$$

where $\mathbf{i}$, $\mathbf{v}$ are $n \times 1$ vectors of the port current and voltage vectors and G is the real constant non-singular admittance matrix. A special case is the three-port circulator, a lossless non-reciprocal device symbolized in Fig. 5.25. With matched termination, power entering port 1 is completely transferred to port 2 with no transfer to port 3. Similarly, power entering port 2 is completely transferred to port 3 with no transfer to port 1, while power entering at port 3 is completely transferred to port 1. The arrow indicates the direction of circulation.

The essential circuit features of the three-port circulator may be explained with the aid of Fig. 5.26. Assuming ideal differential-input amplifiers of gain A, a signal V_1 causes zero output at V_2 (since equal signals are presented to the inputs) and hence no signal will appear at V_4. If a signal is now applied at V_4, a

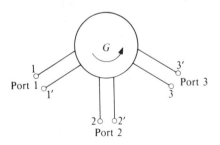

Fig. 5.25. Symbol for three-port circulator

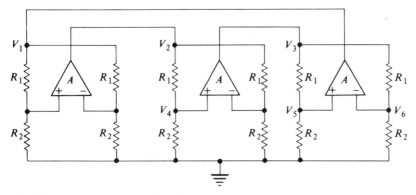

Fig. 5.26. Three-stage terminated circulator

signal appears at V_3 given by
$$V_3 = AV_4$$
and, as a result,
$$V_5 = AV_4 \left(\frac{R_2}{R_1 + R_2} \right)$$

However, since $V_5 = V_6$, no signal is transferred round the loop to V_1. The result is the same whichever ports are considered and the device thus exhibits circulator properties.

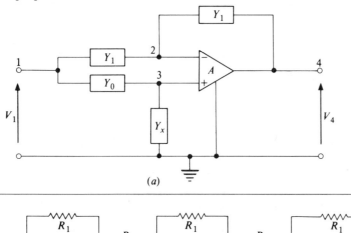

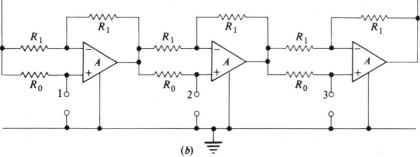

Fig. 5.27. Three-port circulator. (a) Simplified single stage for circulator; (b) practical circuit

Table 5.8 Matrices for ideal active elements

Device	[a]	[g]	[h]	[y]	[z]
VCVS	$\begin{bmatrix} 1/K & 0 \\ 0 & 0 \end{bmatrix}$	$\begin{bmatrix} 0 & 0 \\ K & 0 \end{bmatrix}$			$\begin{bmatrix} 0 & 1/K \\ 0 & 0 \end{bmatrix}$
VCCS	$\begin{bmatrix} 0 & -1/G \\ 0 & 0 \end{bmatrix}$			$\begin{bmatrix} 0 & 0 \\ -G & 0 \end{bmatrix}$	
CCVS	$\begin{bmatrix} 0 & 0 \\ 1/R & 0 \end{bmatrix}$				$\begin{bmatrix} 0 & 0 \\ R & 0 \end{bmatrix}$
CCCS	$\begin{bmatrix} 0 & 0 \\ 0 & -1/K \end{bmatrix}$		$\begin{bmatrix} 0 & 0 \\ -K & 0 \end{bmatrix}$		
VNIC	$\begin{bmatrix} -k & 0 \\ 0 & 1 \end{bmatrix}$	$\begin{bmatrix} 0 & -1 \\ -1/k & 0 \end{bmatrix}$	$\begin{bmatrix} 0 & -k \\ -1 & 0 \end{bmatrix}$		

ACTIVE ELEMENTS

CNIC	$\begin{bmatrix} 1 & 0 \\ 0 & -1/k \end{bmatrix}$	$\begin{bmatrix} 0 & 1/k \\ 1 & 0 \end{bmatrix}$	$\begin{bmatrix} 0 & 1 \\ k & 0 \end{bmatrix}$
IT	$\begin{bmatrix} \pm n & 0 \\ 0 & \pm 1/n \end{bmatrix}$	$\begin{bmatrix} 0 & \mp 1/n \\ \pm 1/n & 0 \end{bmatrix}$	$\begin{bmatrix} 0 & \pm n \\ \mp n & 0 \end{bmatrix}$
VGIC	$\begin{bmatrix} \pm k(s) & 0 \\ 0 & 1 \end{bmatrix}$	$\begin{bmatrix} 0 & -1 \\ \pm 1/k(s) & 0 \end{bmatrix}$	$\begin{bmatrix} 0 & \pm k(s) \\ -1 & 0 \end{bmatrix}$
CGIC	$\begin{bmatrix} 1 & 0 \\ 0 & \pm 1/k(s) \end{bmatrix}$	$\begin{bmatrix} 0 & \mp 1/k(s) \\ 1 & 0 \end{bmatrix}$	$\begin{bmatrix} 0 & 1 \\ \mp k(s) & 0 \end{bmatrix}$
NII	$\begin{bmatrix} 0 & \pm 1/G \\ \mp G & 0 \end{bmatrix}$		$\begin{bmatrix} 0 & \mp G \\ \mp G & 0 \end{bmatrix}$	$\begin{bmatrix} 0 & 0 \\ \mp 1/G & \mp 1/G \end{bmatrix}$
PII	$\begin{bmatrix} 0 & \pm 1/G \\ \pm G & 0 \end{bmatrix}$		$\begin{bmatrix} 0 & \pm G \\ \mp G & 0 \end{bmatrix}$	$\begin{bmatrix} 0 & 0 \\ \pm 1/G & \mp 1/G \end{bmatrix}$

A simplified version of the individual stage[18] is shown in Fig. 5.27(a), for which the nodal admittance matrix is

$$[Y] = \begin{array}{c|cccc} & 1 & 2 & 3 & 4 \\ \hline 1 & Y_1 + Y_0 & -Y_1 & -Y_0 & 0 \\ 2 & -Y_1 & 2Y_1 & 0 & -Y_1 \\ 3 & -Y_0 & 0 & Y_0 + Y_x & 0 \\ 4 & -Y_1 & AY_{44} & -AY_{44} & Y_{44} + Y_1 \end{array}$$

Pivotal condensation about nodes 2 and 3 leads to

$$\frac{V_4}{V_1} = \frac{-Y_{41}}{Y_{44}} = \frac{\left(Y_1 + \dfrac{Y_0 A Y_{44}}{Y_0 + Y_x} - \dfrac{Y_1 A Y_{44}}{2Y_1}\right)}{\left(Y_{44} + Y_1 + \dfrac{Y_1 A Y_{44}}{2Y_1}\right)}$$

and as $A \to \infty$,

$$\frac{V_4}{V_1} = 2\left\{\frac{Y_0}{Y_0 + Y_x} - \frac{1}{2}\right\} \qquad (5.39)$$

from which it can be seen that the gain is

(a) zero if $Y_x = Y_0$
(b) unity if $Y_x = 0$

The stage is thus capable of signal transfer or rejection, as required. A practical implementation of the three-port circulator[18] using the above stage as a building block is shown in Fig. 5.27(b). The admittance matrix can be shown to be

$$[Y] = G \begin{bmatrix} 0 & 1 & -1 \\ -1 & 0 & 1 \\ 1 & -1 & 0 \end{bmatrix} \qquad (5.40)$$

which is identical to the indefinite admittance matrix of the two-port gyrator.

5.5 Perspective

For ease of reference, the matrix descriptions of the ideal forms of two-port active elements discussed in this chapter are combined to form Table 5.8. The relationship between the various elements is illustrated in Fig. 5.28. It can be seen that all controlled sources have three null elements in their transmission matrices, the OA being a special case of the VCVS when the gain tends to

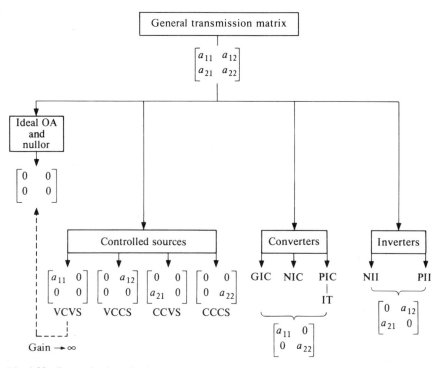

Fig. 5.28. Categorization of active elements according to form of transmission matrix

infinity. Furthermore, the converters have null off-diagonal elements and the inverters have null diagonal elements. The nullor is listed with the ideal OA, while it is clear from earlier sections that pathological elements may be used as the link between the various active elements.

References

1. H. J. Carlin, 'Singular network elements', *IEEE Trans.*, **CT-11**, 67–72, 1964.
2. J. G. Graeme, G. E. Tobey, and L. P. Huelsman, *Operational Amplifiers, Design and Application*, McGraw-Hill, New York, 1971.
3. J. H. Huijsing and J. de Korte, 'Monolithic nullor—a universal active network element', *IEEE Journal*, **SC-12** (1), 59–64, 1977.
4. J. I. Sewell and F. W. Stephenson, 'Matrix tables for the generalized 5-terminal amplifier', *Radio Electron. Engr*, **37** (4), 247–255, 1969.
5. S. K. Mitra, *Analysis and Synthesis of Linear Active Networks*, Wiley, New York, 1969, Chapter 2.
6. A. B. Grebene, *Analog Integrated Circuit Design*, Van Nostrand Reinhold, New York, 1972.
7. M. Bialko and R. W. Newcomb, 'Generation of all finite linear circuits using the integrated DVCCS', *IEEE Trans.*, **CT-18**, 733–736, 1971.

8. H. A. Whitlinger, 'Applications of the CA3080 and CA3080A high-performance operational transconductance amplifier', Application Note ICAN-6668, RCA Solid State Division, Somerville, N.J. 08876, USA, Sept., 1974.
9. D. Hilberman et al. 'A review of active filter terminology, technology and design', IEEE, CAS Network Applications and Standards Committee, Active Filters Sub-Committee Report, 1975.
10. B. D. H. Tellegen, 'The gyrator as a new network element', *Phillips Res. Rept*, **3**, 81–101, 1948.
11. J. G. Linvill, '*RC* active filters', *Proc. IRE*, **12**, 555–564, 1954.
12. A. Antoniou, 'Realization of gyrators using operational amplifiers and their use in *RC*-active-network synthesis', *Proc. IEE*, **116** (11), 1838–1850, 1969.
13. R. G. Hove and C. A. Kleingartner, 'Silicon monolithic gyrator using FETs', *Wescon Tech. Pap.*, pt. 5, Chap. 4'3, pp. 1–5, 1969.
14. H. J. Orchard and A. N. Willson, 'New active gyrator circuit', *Electron. Letters*, **10** (13), 261–262, 1974.
15. J. H. Brodie and R. S. Cocker, 'The active transformer', *Proc. IEEE*, **54**, 1125–1127, 1966.
16. R. H. S. Riordan, 'Simulated inductors using differential amplifiers', *Electron. Letters*, **3** (2), 50–51, 1967.
17. L. T. Bruton, 'Network transfer functions using the concept of frequency-dependent negative resistance', *IEEE Trans.*, **CT-16** (3), 406–408, 1969.
18. J. M. Rollett and P. E. Greenaway, 'Directly coupled active circulators', *Electron. Letters*, **4** (26), 579–580, 1968.

6. Controlled-source realizations

The controlled sources introduced in the previous chapter have active parameters which can be closely adjusted by readily trimmed gain-controlling resistors. The presence of active elements therefore presents less of a problem than was the case in the pre-microelectronic era. Furthermore, it allows consideration of filter systems employing more than one amplifier with a view to achieving improved sensitivity together with fewer and less stringent specifications on the passive elements. In the succeeding sections, controlled sources are applied to the realization of filter transfer functions. The VCVS, in particular, is especially convenient when used in the unity or low-gain mode and can be extended to multiple and mixed-source realizations.

6.1 Single-source methods

It is convenient to limit discussion to the case of the VCVS, but the ensuing methods could equally well be developed for any of the basic controlled sources described in Chapter 5.

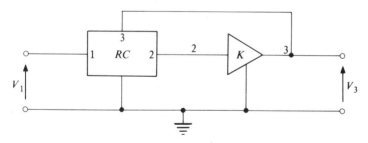

Fig. 6.1. Generalized structure using a single controlled source

Consider the system in Fig. 6.1, in which an ideal VCVS of finite gain K is combined with a four-terminal RC network. Nodal analysis and subsequent application of the constraint $V_3 = KV_2$ (Appendix 3) yields

$$\begin{bmatrix} I_1 \\ I_2 \end{bmatrix} = \begin{bmatrix} y_{11} & y_{12} + Ky_{13} \\ y_{21} & Y_{22} + Ky_{23} \end{bmatrix} \begin{bmatrix} V_1 \\ V_2 \end{bmatrix}$$

where the y-parameters refer to the passive RC network. The voltage transfer ratio is obtained as

$$T = \frac{V_3}{V_1} = K\frac{V_2}{V_1} = \frac{-Ky_{21}}{y_{22} + Ky_{23}} \qquad (6.1)$$

The zeros of the transfer function are unrestricted, being determined by the zeros of an RC transfer admittance. Note the possibility of complex poles since the denominator of Eq. (6.1) may be formed as the difference of two RC immittance functions, allocated as y_{22} and $-Ky_{23}$ respectively. To synthesize a function $T = N(s)/D(s)$ requires the choice of a divisor polynomial

$$Q(s) = \prod_{i-1}^{n-1}(s + \sigma_i) \qquad (6.2)$$

when $D(s)$ is nth order and where the σ_i are real and positive. Thus, in the case of a second-order bandpass function

$$T = \frac{N(s)/Q(s)}{D(s)/Q(s)} = \frac{\dfrac{Hs}{(s + \sigma_1)}}{\left\{s + \dfrac{b_0}{\sigma_1} + \dfrac{b_1 s}{(s + \sigma_1)}\right\} - \dfrac{(\sigma_1 + b_0/\sigma_1)s}{(s + \sigma_1)}}$$

and, with the aid of Eq. (6.1), the y parameters are identified as

$$y_{21} = \frac{-(H/K)s}{(s + \sigma_1)}, \quad y_{22} = \left(s + \frac{b_0}{\sigma_1} + \frac{b_1 s}{s + \sigma_1}\right),$$

and

$$y_{23} = \frac{-(\sigma_1 + b_0/\sigma_1)(s/K)}{(s + \sigma_1)}$$

An RC network having the required form may be obtained by the methods of Chapter 3 and is shown embedded in the structure[1] of Fig. 6.2, for which

$$\frac{V_3}{V_1} = \frac{s\left(\dfrac{KG_1}{C_1}\right)}{s^2 + s\left\{\dfrac{G_2}{C_1}(1 - K) + \dfrac{(G_1 + G_3)}{C_1} + \dfrac{G_3}{C_2}\right\} + \dfrac{G_3(G_1 + G_2)}{C_1 C_2}} \qquad (6.3)$$

Other transfer functions may be synthesized in a similar fashion. However, division of the RC network in Fig. 6.1 yields a system from which a number of

CONTROLLED-SOURCE REALIZATIONS

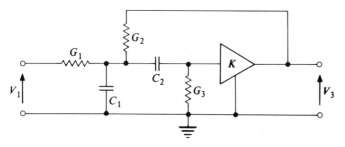

Fig. 6.2. Bandpass filter using positive-gain VCVS

familiar structures are readily derived. The resulting arrangement, shown in Fig. 6.3, yields the following matrix equation for the passive section:

$$\begin{bmatrix} I_1 \\ I_2 \\ I_3 \end{bmatrix} = \begin{bmatrix} (y_{11a} + y_{11b}) & (y_{12a} + y_{12b}) & y_{13a} \\ (y_{21a} + y_{21b}) & (y_{22a} + y_{22b} + Y_a + Y_b) & (y_{23a} - Y_a) \\ y_{31a} & (y_{32a} - Y_a) & (y_{33a} + Y_a) \end{bmatrix} \begin{bmatrix} V_1 \\ V_2 \\ V_3 \end{bmatrix}$$

Substitution of the new matrix elements into Eq. (6.1) gives the following voltage transfer ratio for Fig. 6.3:

$$T = \frac{-K(y_{21a} + y_{21b})}{y_{22a} + y_{22b} + Y_a + Y_b + K(y_{23a} - Y_a)} \qquad (6.4)$$

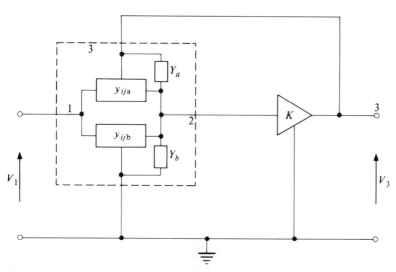

Fig. 6.3. General structure showing modified *RC* network

ACTIVE FILTERS FOR COMMUNICATIONS AND INSTRUMENTATION

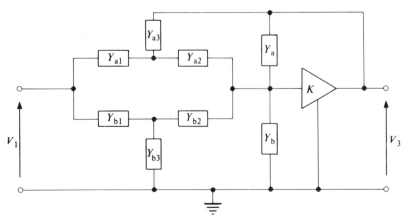

Fig. 6.4. General structure using T networks to realize y_{ija} and y_{ijb}

The use of simple T-networks to realize y_{ija} and y_{ijb} results in the circuit of Fig. 6.4. By the methods of Chapter 3, the y-parameters for the T-networks are

$$y_{21a} = \frac{-Y_{a1}Y_{a2}}{\Sigma y_a}, \quad y_{22a} = \frac{Y_{a2}(Y_{a1} + Y_{a3})}{\Sigma y_a}, \quad y_{23a} = \frac{-Y_{a2}Y_{a3}}{\Sigma y_a}$$

$$y_{21b} = \frac{-Y_{b1}Y_{b2}}{\Sigma y_b}, \quad y_{22b} = \frac{Y_{b2}(Y_{b1} + Y_{b3})}{\Sigma y_b}, \quad y_{23b} = \frac{-Y_{b2}Y_{b3}}{\Sigma y_b}$$

where

$$\Sigma y_a = (Y_{a1} + Y_{a2} + Y_{a3}) \quad \text{and} \quad \Sigma y_b = (Y_{b1} + Y_{b2} + Y_{b3})$$

Substitution into Eq. (6.4) gives

$$T = \frac{K\left\{\dfrac{Y_{a1}Y_{a2}}{\Sigma y_a} + \dfrac{Y_{b1}Y_{b2}}{\Sigma y_b}\right\}}{\left\{\dfrac{Y_{a2}Y_{a3}}{\Sigma y_a}(1-K) + \dfrac{Y_{a1}Y_{a2}}{\Sigma y_a} + \dfrac{Y_{b2}(Y_{b1} + Y_{b3})}{\Sigma y_b} + Y_b + Y_a(1-K)\right\}} \quad (6.5)$$

The system is capable of producing all basic forms of rational transfer function including allpass functions, and Eq. (6.5) constitutes the starting point for all networks discussed in this section.

6.1.1 Non-finite transmission zeros

The transfer ratio for this class of filter may be obtained by appropriate choice of the admittances in Eq. (6.5). Subsequent design proceeds by coefficient matching. If $Y_{b1} = Y_{b2} = Y_{b3} = Y_a = 0$, the modified transfer ratio is given by

$$T = \frac{KY_{a1}Y_{a2}}{Y_{a2}Y_{a3}(1-K) + Y_{a1}Y_{a2} + Y_b\Sigma y_a} \quad (6.6)$$

Table 6.1 Single-source realization of non-finite zeros

Circuit no.	RC structure	Voltage transfer function	Minimum gain constant K_O	Maximum gain constant K_s
1.	(RC structure with G_1, C_1, G_2, C_2)	$\dfrac{\left(\dfrac{KG_1G_2}{C_1C_2}\right)}{s^2 + s\left(\dfrac{G_2}{C_2} + \dfrac{G_1+G_2}{C_1} - \dfrac{KG_2}{C_2}\right) + \dfrac{G_1G_2}{C_1C_2}}$	$\dfrac{C_2}{C_1} + \left(1 - \sqrt{\dfrac{G_1C_2}{G_2C_1}}\right)^2$	$1 + \dfrac{G_1C_2}{G_2C_1} + \dfrac{C_2}{C_1}$
2.	(RC structure)	$\dfrac{Ks^2}{s^2 + s\left(\dfrac{G_2}{C_2} + \dfrac{G_2}{C_1} + \dfrac{G_1}{C_1}(1-K)\right) + \dfrac{G_1G_2}{C_1C_2}}$	$\dfrac{G_2}{G_1} + \left(1 - \sqrt{\dfrac{G_2C_1}{G_1C_2}}\right)^2$	$1 + \dfrac{C_1G_2}{C_2G_1} + \dfrac{G_2}{G_1}$
3.	(RC structure)	$\dfrac{s\left(\dfrac{KG_2}{C_2}\right)}{s^2 + s\left(\dfrac{G_2}{C_2} + \dfrac{G_1+G_2}{C_1}\right) + (1-K)\dfrac{G_1G_2}{C_1C_2}}$	$1 - \dfrac{1}{4}\left(\sqrt{\dfrac{G_2C_1}{G_1C_2}} + \sqrt{\dfrac{G_1C_2}{G_2C_1}} + \sqrt{\dfrac{G_2C_2}{G_1C_1}}\right)^2$	1
4.	(RC structure)	$\dfrac{s\left(\dfrac{KG_1}{(1-K)C_1}\right)}{s^2 + \dfrac{s}{(1-K)}\left(\dfrac{G_1}{C_1}\right)\left(\dfrac{G_2}{C_1} + \dfrac{G_2}{C_2}\right) + \dfrac{G_1G_2}{(1-K)C_1C_2}}$	$1 - \dfrac{1}{4}\left(\sqrt{\dfrac{C_2G_1}{C_1G_2}} + \sqrt{\dfrac{C_1G_2}{C_2G_1}} + \sqrt{\dfrac{G_2C_2}{G_1C_1}}\right)^2$	1

$K > 0$ for circuits 1 and 2, $K < 1$ for circuits 3 and 4.

The simplest possibility for component selection which allows the realization of a second-order function is the replacement of each admittance in Eq. (6.6) by a single passive element. This leads to the structures originally proposed by Sallen and Key.[2]

The resulting RC networks for the realization of low-, high-, and bandpass structures are presented in Table 6.1, together with their associated voltage transfer functions. Of the six possible combinations of two resistors and two capacitors, only the four shown in Table 6.1 yield second-order functions.

To be practically useful, the networks must each be capable of realizing complex poles, a requirement met (in terms of the basic second-order function of Eq. (4.6)) if $b_1^2 < 4b_2 b_0$ and hence if

$$Q = \frac{\sqrt{b_2 b_0}}{b_1} > 0.5$$

The attendant constraint on K for Circuit No. 1 is determined as

$$Q = \frac{\sqrt{\frac{G_1 G_2}{C_1 C_2}}}{\left\{\frac{(G_1 + G_2)}{C_1} + \frac{G_2}{C_2}(1 - K)\right\}} > 0.5 \qquad (6.7)$$

Rearranging the inequality and completing the square gives a lower limit on K as

$$K = K_Q = \frac{C_2}{C_1} + \left(1 - \sqrt{\frac{G_1 C_2}{G_2 C_1}}\right)^2 \qquad (6.8)$$

To produce complex roots, K must be positive and of magnitude greater than K_Q. In addition to realizing complex roots, the circuits must also remain stable. This requires that all denominator coefficients have the same sign. In the case of Circuit No. 1,

$$\frac{G_2}{C_2} + \frac{G_1 + G_2}{C_1} > \frac{K G_2}{C_2}$$

leading to

$$K < K_s = \left(1 + \frac{G_1 C_2}{G_2 C_1} + \frac{C_2}{C_1}\right) \qquad (6.9)$$

From Eqs. (6.8) and (6.9),

$$K_s = K_Q + 2\sqrt{\frac{G_1 C_2}{G_2 C_1}}$$

CONTROLLED-SOURCE REALIZATIONS

so that

$$K_s > K > K_Q > 0 \tag{6.10}$$

This result is also valid for the dual Circuit No. 2, the values of K_s and K_Q being as listed in Table 6.1. For Circuit Nos. 3 and 4, $K < K_Q$ where, again, the values of K_Q are shown in Table 6.1. By inspection of the transfer functions, stability is assured if $K < 1$. The expressions for K_Q make it clear that certain combinations of component values give rise to a range of sub-unity positive values of K. However, the simplest practical solution is to specify a negative gain for the VCVS. Then, for a sufficiently large $|K|$, stable complex poles are guaranteed. The design method is now illustrated by means of an example.

Example 6.1 Design a second-order lowpass Chebyshev filter having 0.5 dB passband ripple, a passband gain of 20 dB and a cut-off frequency of 10 kHz.

SOLUTION The normalized transfer function is (Table 2.1)

$$T = \frac{H}{s^2 + 1.426s + 1.516}$$

A passband gain of 20 dB is equivalent to a modular gain of 10, corresponding to $H = 15.16$. Design proceeds by the method of coefficient matching, as outlined in Sec. 3.4.3. Equating coefficients for Circuit No. 1 of Table 6.1 yields

$$\frac{KG_1 G_2}{C_1 C_2} = 15.16 \tag{6.11a}$$

$$\frac{G_1 G_2}{C_1 C_2} = 1.516 \tag{6.11b}$$

$$\frac{G_2}{C_2} + \frac{G_1 + G_2}{C_1} - \frac{KG_2}{C_2} = 1.426 \tag{6.11c}$$

From Eqs. (6.11a, b), $K = 10$. The remaining two equations and four unknowns indicate freedom of choice for two elements. For convenience, let $C_1 = C_2 = 1$, then the equations reduce to

$$G_1 G_2 = 1.516 \tag{6.12a}$$

$$G_1 - 8G_2 = 1.426 \tag{6.12b}$$

from which the only solution having positive values is

$$G_1 = 4.268 \quad \text{and} \quad G_2 = 0.355$$

The normalized design is shown in Fig. 6.5(a). The frequency denormalization factor is specified as $2\pi \times 10^4$ and, choosing an impedance denormalization factor (Appendix 1) of 10^4, yields the following set of practical component values:

$$C_1 = C_2 = 0.0016 \text{ μF}$$

$$R_1 = \frac{1}{G_1} = 2.34 \text{ k}\Omega \quad \text{and} \quad R_2 = \frac{1}{G_2} = 28.17 \text{ k}\Omega$$

Two extra resistors are required to establish the value of K. The circuit for the VCVS is shown as the non-inverting arrangement in Table 5.4.

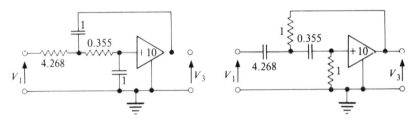

Fig. 6.5. Realization of second-order 0.5 dB ripple Chebyshev characteristic. Element values in siemens and farads. (a) Normalized lowpass filter; (b) highpass dual of (a)

The highpass filter (Circuit No. 2) may be designed in the same manner as the lowpass, the transfer function (if not directly available) being readily obtained by the lowpass–highpass frequency transformation of Chapter 2. Alternatively, the highpass filter may be directly formed from the lowpass by invoking the Principle of Duality (Sec 3.2.3), interchanging capacitors and resistors, as evidenced in Table 6.1 and Fig. 6.5(b), which is the highpass dual of Fig. 6.5(a). The two bandpass sections in Table 6.1 are also duals of each other and may be designed in a similar manner to the lowpass case. An equal-resistor and equal-capacitor solution is possible if the passband gain is relaxed to $3 - \frac{1}{Q}$, as in Table 6.5.

Of the three designs for bandpass filters presented, one requires a positive-gain VCVS (Fig. 6.2) and the other two a negative-gain VCVS (Circuits 3 and 4 in Table 6.1). Important differences between the positive- and negative-gain designs become evident from a study of the Q factor and stability limits. For the positive-gain system, the Q factor is given by

$$Q = \frac{\sqrt{\frac{(G_1 + G_2)C_2}{G_3 C_1}}}{1 + \frac{(G_1 + G_3)C_2}{G_3 C_1} + \frac{G_2 C_2}{G_3 C_1}(1 - K)} \tag{6.13}$$

being enhanced (as in Ex. 6.1) by the subtractive effect on the denominator as K is increased. However, there is a limit to the value of K if stability is to be maintained:

$$K < K_s = 1 + \frac{(G_1 + G_3)}{G_2} + \frac{G_3 C_1}{G_2 C_2} \qquad (6.14)$$

This implies that a demand for high Q factors is achieved at the expense of increased susceptibility to instability when a positive-gain VCVS is used. A different result is achieved when the VCVS has negative gain, the two bandpass circuits of Table 6.1 having identical selectivities:

$$Q = \frac{\sqrt{\dfrac{(1-K)G_1 G_2}{C_1 C_2}}}{\left\{\dfrac{(G_1 + G_2)}{C_1} + \dfrac{G_2}{C_2}\right\}} \qquad (6.15)$$

In this case, the system is absolutely stable, but the sign equality of the denominator terms indicates that there is no possibility of Q enhancement except by employing high values of gain or wide component spreads.

6.1.2 Imaginary transmission zeros

Finite transmission zeros, as required in band-elimination filters, can be realized by superimposing the lowpass and highpass RC structures of Table 6.1 as in Fig. 6.6 to produce a notch function

$$T = \frac{H(s^2 + a_0)}{s^2 + b_1 s + b_0} \qquad (6.16)$$

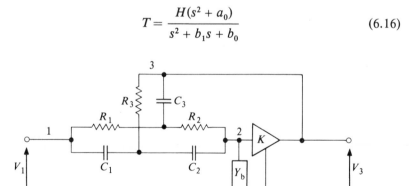

Fig. 6.6. Floating twin-T realization of notch function

This can be recognized in Fig. 6.3 as using a null-adjusted passive RC twin-T network for y_{ija}, giving the transfer ratio

$$\frac{V_3}{V_1} = \frac{-y_{21a}}{(y_{22a} + Y_b)/K - (y_{21a} + y_{22a})} \tag{6.17}$$

Substituting parameters for the symmetrical twin-T of Chapter 3 gives the transfer function

$$T = \frac{\dfrac{K}{(1+2\sigma)}\left\{\left(\dfrac{s}{\omega_0}\right)^2 + 1\right\}}{\left(\dfrac{s}{\omega_0}\right)^2 + \dfrac{4}{\tau(1+2\sigma)}\left[1 + \left(\dfrac{v+\sigma}{2}\right) - K\right]\left(\dfrac{s}{\omega_0}\right) + \left(\dfrac{1+2v}{1+2\sigma}\right)} \tag{6.18}$$

where

$$\sigma = C_4/C, \quad v = R/R_4 = G_4/G, \quad C_1 = C_2 = C = C_3/2,$$
$$R_1 = R_2 = R = 2R_3, \quad \omega_0 = 1/RC, \quad \text{and} \quad Y_b = (G_4 + sC_4)$$

For a symmetrical response, $\sigma = v$, in which case the network may be designed by choosing τ for a given notch frequency, K for passband gain, then v for selectivity.

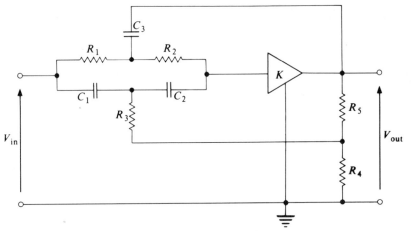

Fig. 6.7. Controlled positive-feedback Lim notch filter circuit

A variant of the floating twin-T circuit with controlled positive feedback[3,4] is shown in Fig. 6.7. Again assuming symmetrical null adjustment, the transfer function is

$$\frac{V_{out}}{V_{in}} = \frac{\dfrac{1}{K}\left\{\left(\dfrac{s}{\omega_0}\right)^2 + 1\right\}}{\left(\dfrac{s}{\omega_0}\right)^2 + 4\left(1 - \dfrac{1}{2K}\cdot\dfrac{1+2q}{1+q}\right)\left(\dfrac{s}{\omega_0}\right) + 1} \tag{6.19}$$

where

$$q = \frac{R_4}{R_5} \quad \text{and} \quad \omega_0 = 1/2C\left(R_3 + \frac{R_4 R_5}{R_4 + R_5}\right)$$

Obviously, resistances R_4 and R_5 control the Q, but they also affect the centre frequency ω_0, thereby complicating tuning.

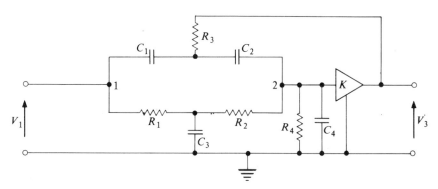

Fig. 6.8. Grounded twin-T network realizing complex zeros

Suitable notch networks can also be derived from the general grounded-T structure of Fig. 6.4. Using single-element replacements for all admittances except Y_b, the required transfer function of Eq. (6.16) can be realized by the following methods:

(a) Setting $Y_a = 0$. An example[1] incorporating resistive feedback and two grounded capacitors is shown in Fig. 6.8. Following the evaluation of preliminary design parameters $\mu = (2.5\sqrt{b_0} - b_0)$, $\psi = (2.5\sqrt{b_0} - 1)/b_0$, $\delta = a_0/(a_0 + b_0)$ and $\rho = b_0/(a_0 + b_0)$, the element values may be obtained as

$$C_1 = \rho\psi/(\psi - 1), \quad C_2 = \rho\psi, \quad C_3 = \delta\mu^2/b_0(\mu - 1), \quad C_4 = \delta$$
$$G_1 = \delta\mu/(\mu - 1), \quad G_2 = \delta\mu, \quad G_3 = \rho b_0 \psi^2/(\psi - 1), \quad G_4 = \rho$$
$$K = \rho b_0 \psi/(2.5\sqrt{b_0} - b_1), \text{ and the gain constant, } H = K\rho.$$

(b) Setting $Y_b = 0$ in Fig. 6.4 and letting $\sum y_a = \sum y_b$ so that Eq. (6.5) reduces to

$$\frac{V_3}{V_1} = \frac{K(Y_{a1}Y_{a2} + Y_{b1}Y_{b2})}{Y_{a1}Y_{a2} + Y_{b2}(Y_{b1} + Y_{b3}) + (1 - K)[Y_a(Y_{a1} + Y_{a2} + Y_{a3}) + Y_{a2}Y_{a3}]} \quad (6.20)$$

One realization of Eq. (6.20) is the selection[5] $Y_{b1} = sC_1$, $Y_{b2} = sC_2$, $Y_{b3} = (G_1 + G_2)$, $Y_{a1} = G_1$, $Y_{a2} = G_2$, $Y_{a3} = s(C_1 + C_2)$, and $Y_a = G$, which yields

$$\frac{V_3}{V_1} = \frac{K(s^2 C_1 C_2 + G_1 G_2)}{s^2 C_1 C_2 + s[C_2(G_1 + G_2) + (C_1 + C_2)(G_2 + G) \times (1 - K)] + [G_1 G_2 + G(G_1 + G_2)(1 - K)]} \qquad (6.21)$$

Design may now proceed by coefficient matching, as illustrated in the following example.

Example 6.2 Design a second-order filter with transfer function

$$\frac{V_3}{V_1} = \frac{H(s^2 + 3.414)}{s^2 + 0.836s + 1.224}$$

SOLUTION By comparison with Eq. (6.21),

$$C_1 C_2 = 1 \qquad (6.22a)$$
$$G_1 G_2 = 3.414 \qquad (6.22b)$$
$$C_2(G_1 + G_2) + (C_1 + C_2)(G_2 + G)(1 - K) = 0.836 \qquad (6.22c)$$
$$G_1 G_2 + G(G_1 + G_2)(1 - K) = 1.224 \qquad (6.22d)$$

Since there are six unknowns and only four equations, some element selection is necessary. For ease of fabrication, set $C_1 = C_2 = 1$ and, from the modified Eq. (6.22c), let

$$G_1 + G_2 = 0.836(n + 1)$$

Then, solving

$$G_1 = \frac{(n + 1) \pm \sqrt{(n + 1)^2 (0.836)^2 - 4(3.414)}}{2}$$

which will always have two positive-real roots if $(n + 1) > 4.42$. Choosing $n = 4$ gives $G_1 = 3.067$ (or 1.113) and $G_2 = 1.113$ (or 3.067). Selecting the first value in each case gives the final circuit shown in Fig. 6.9, with element values as indicated.

The relative merits of the basic configurations described in this section for realizing notch responses with imaginary zeros can be compared in terms of selectivity, stability, and sensitivity. Inspection of Eq. (6.18) indicates that, for

CONTROLLED-SOURCE REALIZATIONS

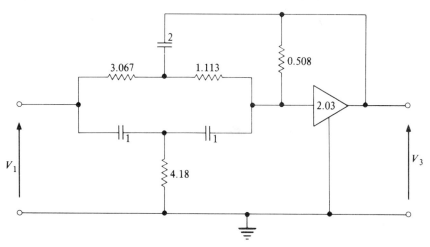

Fig. 6.9. Realization of elliptic function of Ex. 6.2. Element values in siemens and farads

the symmetrical response case, the floating twin-T realization has a selectivity

$$Q = \frac{1}{4} \cdot \left(\frac{1 + 2\sigma}{1 + \sigma - K} \right) \tag{6.23}$$

while stability sets an upper limit to amplifier gain according to

$$K < K_s = 1 + \sigma \tag{6.24}$$

which is a severe restriction, especially in the absence of G_4 and C_4.

With the controlled-feedback floating twin-T, Eq. (6.19) gives the stability limit

$$K < K_s = \frac{1}{2} \cdot \left(\frac{2q + 1}{q + 1} \right) \tag{6.25}$$

For high selectivity, this dictates the choice of unity gain, obtained by direct negative feedback of an operational amplifier (Chapter 5), and implying that

$$Q = \frac{q}{2} \tag{6.26}$$

For the symmetrical-response grounded twin-T configuration of Fig. 6.8,

$$Q = \frac{1}{4} \cdot \left(\frac{1 + 2\sigma}{1 + \sigma - K/2} \right) \tag{6.27}$$

with stability limit

$$K < K_s = 2(1 + \sigma) \tag{6.28}$$

Therefore, it exhibits a wider range of stable operation than the other circuits. However, turning to the Q sensitivities, for the floating arrangements of Fig. 6.6,

$$S_K^Q = \frac{4K}{1+2\sigma} Q \qquad (6.29)$$

which is the same as for Fig. 6.7 at high Q and $K = 1$, in the non-shunted case. For the grounded arrangement of Fig. 6.8, it is half this value. Such high sensitivities are characteristic of realizations relying on the null adjustment of a twin-T section.

6.2 Dual-source methods

This section discusses methods of introducing a second source into the general system of Fig. 6.1.

6.2.1 Active isolation

The resonator blocks discussed in Sec. 6.1.1 can be represented by the general system shown with its signal-flow graph in Fig. 6.10. The complex-zero realizations of Sec. 6.1.2 necessitate introduction of an additional node. Yet another case of the general schematic of Fig. 6.1 is obtained by replacing the passive-RC network by an active RC network, notably by introducing a second VCVS (K_1) at node 2', as in Fig. 6.11. This is attended by several advantages:

(a) As indicated by the modified signal-flow graph of Fig. 6.11(b), component interaction is reduced, thereby leading to simplified design equations.

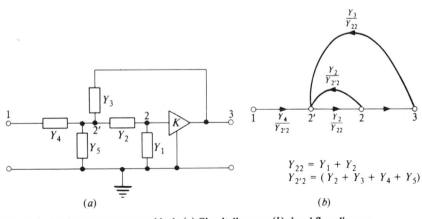

Fig. 6.10. Single-source resonator block. (a) Circuit diagram; (b) signal-flow diagram

CONTROLLED-SOURCE REALIZATIONS

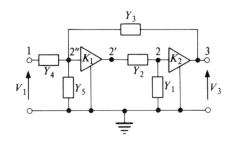

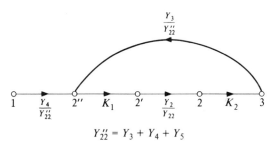

$$Y_{22}'' = Y_3 + Y_4 + Y_5$$

Fig. 6.11. Dual-source resonator block. (*a*) Circuit diagram; (*b*) signal-flow diagram

(b) The buffering effect of the additional VCVS allows narrower component spreads, at the expense of two gain-controlling resistors.

(c) The system gain can be shared between two sources, thereby imposing less severe demands on gain in the negative-feedback case.

(d) More flexibility is obtained in optimizing sensitivity performance (see Sec. 6.4).

Analysing the single-feedback-loop graph of Fig. 6.11(*b*) by Mason's general rule (Appendix 4) with

$$Y_{22} = Y_1 + Y_2 \quad \text{and} \quad Y_{22}'' = Y_3 + Y_4 + Y_5$$

gives the transfer ratio

$$T = \frac{V_3}{V_1} = \frac{KY_2 Y_4}{Y_1(Y_3 + Y_4 + Y_5) + Y_2(Y_4 + Y_5) + Y_2 Y_3(1 - K)} \quad (6.30)$$

where the total system gain $K = K_1 K_2$.

Using the same element allocation as for the single-source resonators, the resulting voltage transfer functions are presented in Table 6.2, the final entry being derived from the 5-element structure of Fig. 6.2. Evaluation of K_s and K_ϱ (in the manner of Sec. 6.1.1) leads to the conclusion that the additional VCVS's should, in all cases, be noninverting. Thus, Circuits 1 and 2 contain two positive-gain VCVS while Circuits 3 and 4 contain one positive- and one negative-gain source. The results are not conclusive for Circuit 5 (for which K_ϱ can be negative) but, since K_s has a form similar to that for Circuits 1 and 2, it is convenient to retain the system as one with overall positive gain.

Table 6.2 Dual-source realization of non-finite transmission zeros

Circuit no.	Voltage transfer function	K_Q	K_s	Total system gain (K)	Q enhancement (Q_2/Q_1)
1.	$\dfrac{\left(\dfrac{KG_1G_2}{C_1C_2}\right)}{s^2 + s\left(\dfrac{G_2}{C_2}(1-K) + \dfrac{G_1}{C_1}\right) + \dfrac{G_1G_2}{C_1C_2}}$	$\left(1 - \sqrt{\dfrac{C_2G_1}{C_1G_2}}\right)^2$	$1 + \dfrac{C_2G_1}{C_1G_2}$	Positive	$\dfrac{1}{1 + \dfrac{G_1}{G_2} + \dfrac{C_1}{C_2}(1-K)}$
2.	$\dfrac{Ks^2}{s^2 + s\left(\dfrac{G_2}{C_2} + \dfrac{G_1}{C_1}(1-K)\right) + \dfrac{G_1G_2}{C_1C_2}}$	$\left(1 - \sqrt{\dfrac{C_1G_2}{C_2G_1}}\right)^2$	$1 + \dfrac{C_1G_2}{C_2G_1}$	Positive	$\dfrac{1}{1 + \dfrac{C_1}{C_2} + \dfrac{G_1}{G_2}(1-K)}$
3.	$\dfrac{s\left(\dfrac{KG_2}{C_2}\right)}{s^2 + s\left(\dfrac{G_1}{C_1} + \dfrac{G_2}{C_2}\right) + (1-K)\dfrac{G_1G_2}{C_1C_2}}$	$1 - \dfrac{1}{4}\left\{\sqrt{\dfrac{C_2G_1}{C_1G_2}} + \sqrt{\dfrac{C_1G_2}{C_2G_1}}\right\}^2$	1	Negative	$\dfrac{1}{1 + \dfrac{G_1}{G_2} + \dfrac{C_1}{C_2}}$
4.	$\dfrac{s\left(\dfrac{KG_1}{(1-K)C_1}\right)}{s^2 + s\dfrac{\left(\dfrac{G_2}{C_2} + \dfrac{G_1}{C_1}\right)}{(1-K)} + \dfrac{G_1G_2}{(1-K)C_1C_2}}$	$1 - \dfrac{1}{4}\left\{\sqrt{\dfrac{C_2G_1}{C_1G_2}} + \sqrt{\dfrac{C_1G_2}{C_2G_1}}\right\}^2$	1	Negative	$\dfrac{1}{1 + \dfrac{C_1}{C_2} + \dfrac{G_1}{G_2}}$
5.	$\dfrac{s\left(\dfrac{KG_1}{C_1}\right) + \dfrac{G_3(G_1+G_2)}{C_1C_2}}{s^2 + s\left(\dfrac{G_3}{C_2} + \dfrac{G_1}{C_1} + \dfrac{G_2}{C_1}(1-K)\right) + \dfrac{G_3(G_1+G_2)}{C_1C_2}}$	$\dfrac{G_1}{G_2}\left(1 - \dfrac{G_3C_1}{G_2C_2}\right) + \left\{1 - \sqrt{\dfrac{G_3C_1}{G_2C_2}\left(1+\dfrac{G_1}{G_2}\right)}\right\}^2$	$1 + \dfrac{G_1}{G_2} + \dfrac{C_1G_3}{C_2G_2}$	Positive	$\dfrac{1}{1 + \dfrac{C_1}{C_2} + \dfrac{G_2}{G_3}(1-K) + \dfrac{G_1}{G_3}}$

$K - KK$

CONTROLLED-SOURCE REALIZATIONS

In order to assess the effect of employing a dual source on selectivity, consider Circuit No. 1 of Table 6.1 and Table 6.2 in turn. The single-source arrangement has

$$Q_1 = \frac{\sqrt{\frac{G_1 G_2}{C_1 C_2}}}{\left\{ \frac{(G_1 + G_2)}{C_1} + \frac{G_2}{C_2}(1 - K) \right\}} \tag{6.31}$$

while for the dual-source system,

$$Q_2 = \frac{\sqrt{\frac{G_1 G_2}{C_1 C_2}}}{\left\{ \frac{G_1}{C_1} + \frac{G_2}{C_2}(1 - K) \right\}} \tag{6.32}$$

leading to the selectivity ratio Q_2/Q_1, as listed in Table 6.2. For stability, the second term of this ratio can never be negative, hence $Q_2/Q_1 > 1$ and selectivity is always enhanced by addition of the second source. Results for the other circuits are presented in Table 6.2 from which the two negative-gain systems are seen to have selectivity ratios independent of K. This useful feature is tempered by the fact that ω_0 sensitivity is finite as compared to zero for the positive-gain systems.

Example 6.3 Realize the second-order lowpass transfer function of Eq. (4.6) by means of circuit No. 3 using

(a) a single VCVS
(b) two VCVS's.

Compare the two realizations with regard to total system gain and component spread.

SOLUTION In each case, design proceeds by coefficient matching. For the single-source system, from Table 6.1, the design equations are

$$(1 - K) \frac{G_1 G_2}{C_1 C_2} = b_0$$

and

$$\frac{G_2}{C_2} + \frac{(G_1 + G_2)}{C_1} = b_1$$

141

which combine to yield

$$\frac{b_0}{b_1^2} = Q^2 = (1-K)\frac{G_1 G_2}{C_1 C_2}\left\{\frac{G_2}{C_2} + \frac{(G_1+G_2)}{C_1}\right\}^{-2}$$

For equal-valued resistors and capacitors, the required amplifier gain is therefore $K = (1 - 9Q^2)$. For the dual-source arrangement, the corresponding coefficient equations yield

$$Q^2 = \frac{b_0}{b_1^2} = (1-K)\frac{G_1 G_2}{C_1 C_2}\left\{\frac{G_1}{C_1} + \frac{G_2}{C_2}\right\}^{-2}$$

indicating a total gain requirement (for equal elements) of $K = (1 - 4Q^2)$. This may be shared between the two amplifiers, e.g., $K_1 = -K_2 = \sqrt{4Q^2 - 1}$.

Thus, the dual-source system exhibits lower individual source gain requirements than the single-source system for equivalent component spreads. It should be noted that an additional advantage of the dual-source system is its higher Q for equivalent component spreads, as can be checked with the aid of Table 6.2.

Since it is relatively simple to realize unity-gain VCVS's which are fairly insensitive to variations in the gains of constituent physical devices (see Chapter 5), it is convenient to use them in filter design. A particular example is the lowpass section shown in Fig. 6.12 which, because of the low system-gain

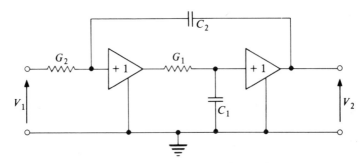

Fig. 6.12. Unity-gain-source filter suitable for use at high frequencies

requirement, is applicable[6] to the realization of pole frequencies up to 3 MHz. Such design may be alleviated by resort to emitter-follower transistor stages since operational amplifiers cannot easily achieve high Q's at these frequencies (see Sec. 9.4). The gain constraint has no effect on centre frequency but, from Table 6.2 (Circuit No. 1), the Q factor is reduced to:

$$Q = \sqrt{\frac{G_2 C_1}{G_1 C_2}} \qquad (6.33)$$

and is thus limited to values set by practically desirable component spreads.

CONTROLLED-SOURCE REALIZATIONS

Highpass realizations can also be obtained with $K_1 = K_2 = 1$, but the resulting elimination of the $Y_2 Y_3$ term (Eq. (6.30)) precludes a four-element bandpass realization. This form of characteristic must therefore be formed from the five-element system (Circuit No. 5, Table 6.2) if voltage followers are used. Other design constraints, largely aimed at easing fabrication problems, are equal-capacitor solutions[7] and those in which either the total resistance sum or the resistance spread is minimized.[8]

6.2.2 RC embedding

The bandpass positive-feedback decomposition outlined in Sec. 4.2.4 for the sensitivity minimization of a second-order function with respect to the gain of

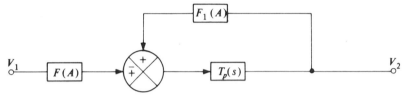

Fig. 6.13. Block diagram for *RC*-embedded bandpass-positive-feedback structures

the active element suggests a dual-source feedback arrangement which leads to a family of *RC*-embedded structures. With $F_2(A) = 0$ and the incorporation of a forward gain dependence $F(A)$, Fig. 4.6 degenerates to the block diagram form of Fig. 6.13. Hence the overall transfer function is given by

$$T(s) = \frac{\mp F(A)}{\dfrac{1}{T_p(s)} - F_1(A)} \qquad (6.34)$$

where $T_p(s)$ is the transfer function of a passive section, $F(A)$ and $F_1(A)$ are functions of the active element, and the sign is determined by the nature of the input connection. This may be represented in network notation by Fig. 6.14, which realizes $T_p(s)$ by means of a three-terminal passive *RC* structure. The

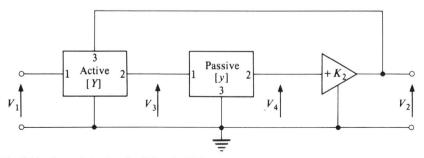

Fig. 6.14. General structure for *RC*-embedded systems

Table 6.3 Positive-feedback active sections and parameters

Active section		Admittance matrix $[Y]$		
Type	Circuit	1	2	3
Summed input	(circuit with amplifier K_1, resistors G_A, G_B, nodes 1, 3, 4, 2)	1: $\dfrac{G_A G_B}{G_A + G_B}$ 2: $\dfrac{-K_1 G_A Y'_{22}}{G_A + G_B}$ 3: $\dfrac{-G_A G_B}{G_A + G_B}$	1: 0 2: Y'_{22} 3: 0	1: $\dfrac{-G_A G_B}{G_A + G_B}$ 2: $\dfrac{-K_1 Y'_{22} G_B}{G_A + G_B}$ 3: $\dfrac{G_A G_B}{G_A + G_B}$
Difference input	(circuit with op-amp A_1, resistors G_A, G_B, nodes 1, 3, 4, 2)	1: $\dfrac{G_A G_B}{G_A + G_B}$ 2: $\dfrac{G_A A_1 Y'_{22}}{G_A + G_B}$ 3: 0	1: $\dfrac{-G_A G_B}{G_A + G_B}$ 2: $Y'_{22} + \dfrac{G_B A_1 Y'_{22}}{G_A + G_B}$ 3: 0	1: 0 2: $-A_1 Y'_{22}$ 3: 0

Y'_{22} is the output admittance of the active element.

latter is fed by an active four-terminal network controlled by both the input and feedback from the output. Unlike designs in the previous section, the active devices 'embed' the RC structure rather than isolate it into subsections. In terms of the external terminal notation, the constraint $V_2 = K_2 V_3$ emanating from Appendix 3 results in the overall matrix equation

$$\begin{bmatrix} I_1 \\ I_3 \\ I_4 \end{bmatrix} = \begin{bmatrix} Y_{11} & Y_{12} & K_2 Y_{22} \\ Y_{21} & Y_{22} + y_{11} & K_2 Y_{23} + y_{12} \\ 0 & y_{21} & y_{22} \end{bmatrix} \begin{bmatrix} V_1 \\ V_3 \\ V_4 \end{bmatrix} \qquad (6.35)$$

where the Y_{ii}, Y_{ij} terms refer to the active section and the y_{ii}, y_{ij} terms to the passive section. Following pivotal condensation about node 3, the voltage transfer ratio becomes

$$T = \frac{K_2 V_4}{V_1} = \frac{K_2 Y_{21} y_{21}}{y_{22}(Y_{22} + y_{11}) - y_{21}(Y_{12} + K_2 Y_{23})} \qquad (6.36)$$

Two implementations of the active block are shown in Table 6.3 alongside the corresponding active $[Y]$ parameters derived by the methods of Chapter 5. Substituting into Eq. (6.36) for the positive-feedback summation input gives

$$T = \frac{-K_1 K_2/(1+q)}{y_{22}/y_{21} + q K_1 K_2/(1+q)} \qquad (6.37)$$

where $q = G_B/G_A$, while, for the difference-input system,

$$T = \frac{\dfrac{K_2}{q + (q+1)/A_1}}{\dfrac{y_{22}}{y_{21}} + \dfrac{K_2}{q/(q+1) + 1/A_1}} \qquad (6.38)$$

Recognizing the voltage transfer function of the passive RC network as $T_p = -y_{21}/y_{22}$, then $F(A)$ and $F_1(A)$ may be identified in the general Eq. (6.34). The positive numerator occurs for the summation input and the negative for the difference input. When T_p has the normalized bandpass form

$$T_p(s) = \frac{s}{s^2 + bs + 1} \qquad (6.39)$$

then Eq. (6.34) becomes

$$T(s) = \frac{\mp F(A)s}{s^2 + [b - F_1(A)]s + 1} \qquad (6.40)$$

Table 6.4 *RC*-embedded bandpass realizations

Type	*RC*-embedded structure	Voltage transfer function
	Circuit diagram	V_2/V_1
Summation-input Twin-T		$\dfrac{4\left(\dfrac{K_1 K_2}{q+1}\right)\left(\dfrac{sC}{G}\right)}{\left(\dfrac{sC}{G}\right)^2 + 4\left[1 - \dfrac{qK_1 K_2}{1+q}\right]\left(\dfrac{sC}{G}\right) + 1}$ $G_1 = G_2 = G_3/2 = G \quad \text{and} \quad C_1 = C_2 = C_3/2 = C$
Difference-input Bridged-T		$\dfrac{-\left[\dfrac{K_2}{q(1+1/A_1)+1/A_1}\right]\left(\dfrac{sC}{G}\right)}{\left(\dfrac{sC}{G}\right)^2 + \left[2+n - \dfrac{K_2}{q/(q+1)+1/A_1}\right]\left(\dfrac{sC}{G}\right) + 1}$ $G_1 = G_2/n \quad \text{and} \quad C_1 = C_2/n$

which is also a bandpass function with a selectivity controlled by the active feedback. The results of Chapter 3 suggest that Eq. (6.39) be realized by RC circuits which are effectively rotated twin-T or bridged-T arrangements.

Two possibilities are illustrated in Table 6.4. The summation-input twin-T circuit[9] with $K_1 = K_2 = 1$ allows single-element control by G_A or G_B of the selectivity $Q = (q + 1)/4$, while at the same time exhibiting unity overall gain at resonance for all values of Q. Using a bridged-T in the same active system gives

$$Q = \frac{1}{2 + n - [q/(q + 1)]K_1 K_2} \qquad (6.41)$$

which is not so conveniently controlled because of the dependence on the resistor and capacitor ratios n of the passive network as well as on q. The same is true of the difference-input bridged-T circuit[10] in Table 6.4, for which

$$Q = \frac{1}{2 + n - (1 + 1/q)K_2} \qquad (6.42)$$

assuming ideal A_1, whereas the same arrangement using a twin-T has

$$Q = \frac{1}{4[1 - (1 + 1/q)K_2]} \qquad (6.43)$$

This indicates single-element control only for $K_2 < 1$, the system being otherwise unstable. Thus, in all cases, non-unity K_2 gives another degree of freedom at the expense of extra control resistors. The design of the two circuits presented in Table 6.4 proceeds by coefficient matching, the high degree of element equality obviating the need for an example. Other realizations using two[11] or more[9] VCVS's have been suggested which combine the advantages of single-element control of Q factor, simple relationships between element values, and readily obtained gain for at least one of the sources.

6.3 Mixed-source methods

Designs in the previous sections have all used the same type of controlled source as the active element. There are occasionally advantages to be gained from methods which utilize controlled sources of differing types in the same network. Three such methods are now to be discussed, each having its own design advantages.

6.3.1 Unity-gain controlled-source realizations

In order to take advantage of the unity-gain source features outlined in Sec. 6.2.1, consider the realization of the general three-terminal cascade structure in

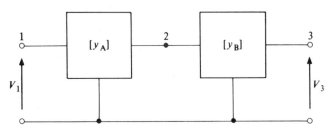

Fig. 6.15. General cascade structure

Fig. 6.15. Analysis and application of the matrix conversions of Appendix 5 gives

$$\frac{V_3}{V_1} = \frac{-y_{21A} Z_{21B}}{y_{22A} Z_{11B} + 1} \quad (6.44)$$

Now, if the input impedance of network B is negligibly small, the transfer ratio reduces to

$$\frac{V_3}{V_1} = -y_{21A} Z_{21B} \quad (6.45)$$

being determined only by transfer immittances. Such conditions may be implemented by the structure[12] of Fig. 6.16, which utilizes a VCVS† and two CCCS's, K_1 and K_3. CCCS$_1$ ensures zero input impedance for network B and

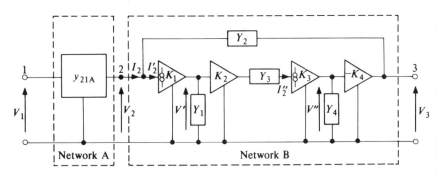

Fig. 6.16. General structure using unity-gain controlled sources

† In Cheng and Stromsoe,[12] these devices are referred to as GVUGA (grounded voltage unit-gain amplifier) and GCUGA (grounded current unity-gain amplifier), respectively. However, in this text it is more convenient to retain the terms VCVS and CCCS in which the gain constants are unity.

the transfer impedance can be evaluated relatively simply. At the input to network B,

$$I'_2 = I_2 + V_3 Y_2 \qquad (6.46a)$$

(since $V_2 = 0$).
Also, at nodes V', V'', and V_3,

$$V' Y_1 = K_1 I'_2 \qquad (6.46b)$$
$$V'' Y_4 = K_3 I''_2 \qquad (6.46c)$$
$$V_3 = -K_4 V'' \qquad (6.46d)$$

and, finally, I''_2 is obtained as

$$I''_2 = K_2 V' Y_3 \qquad (6.46e)$$

Straightforward manipulation of the above equations leads to

$$Z_{21B} = \frac{V_3}{I_2} = \left\{ Y_2 + \frac{Y_1 Y_4}{Y_3 \alpha} \right\}^{-1} \qquad (6.47)$$

where $\alpha = K_1 K_2 K_3 K_4$
Hence, with the aid of Eq. (6.45), the voltage transfer function is

$$T = \frac{V_3}{V_1} = \frac{y_{21A} \alpha}{\{Y_2 \alpha + Y_1 Y_4 / Y_3\}} \qquad (6.48)$$

Synthesis proceeds by means of the summation decomposition (Chapter 4). Writing T as a ratio of two polynomials in s, an arbitrary divisor polynomial, $Q(s)$, may be chosen such that

$$-T = \frac{N(s)}{D(s)} = \frac{N(s)/Q(s)}{D(s)/Q(s)}$$

$D(s)/Q(s)$ must now be expressed as a sum of driving-point RC and RL admittances. Choosing $Y_4 = G_4 = 1$ and $Y_1 = G_1 = 1$,

$$\frac{D(s)}{Q(s)} = \frac{1}{Y_3} + \alpha Y_2 \qquad (6.49)$$

which is of the required form.
Consider the synthesis of a second-order frequency-normalized transfer function

$$-T(s) = \frac{N(s)}{s^2 + b_1 s + 1} \qquad (6.50)$$

ACTIVE FILTERS FOR COMMUNICATIONS AND INSTRUMENTATION

A suitable choice for $Q(s)$ (satisfying Calahan's decomposition requirement of Sec. 4.2.3) is

$$Q(s) = (s + b_1/2) \qquad (6.51)$$

Hence, from Eqs. (6.49) and (6.50),

$$\frac{D(s)}{Q(s)} = s + \frac{b_1}{2} + \frac{1 - (b_1/2)^2}{(s + b_1/2)}$$

Comparing this with Eq. (6.48) allows the identifications

$$Y_2 = s + b_1/2 = sC_2 + G_2 \qquad (6.52a)$$

$$Y_3 = \frac{(s + b_1/2)}{1 - (b_1/2)^2} = sC_3 + G_3 \qquad (6.52b)$$

Substitution of the element values for Y_1, Y_2, Y_3, and Y_4 into Eq. (6.48) yields the general voltage transfer function as

$$T(s) = \frac{V_3}{V_1} = \frac{y_{21A}(sC_3 + G_3)}{s^2 C_2 C_3 + s[C_3 G_2 + C_2 G_3] + [G_2 G_3 + G_1 G_4/\alpha]} \qquad (6.53)$$

The specific form of the realized transfer function now depends upon the choice of network A for which the pole of y_{21A} cancels out the numerator zero of $T_v(s) = (sC_3 + G_3)$. Examples suitable for a selection of transfer functions are presented in Chapter 3.

Example 6.4 Design a second-order bandpass filter having a Q factor of 10 and a centre-band gain of 20 dB.

SOLUTION The normalized transfer function is of the form of Eq. (6.50) with $b = 0.1$ and numerator constant $H = b_1$ (antilog$_{10}$ 1) = 1. The optimum divisor polynomial is $Q(s) = (s + 0.5)$. Select $G_4 = G_1 = 1$ and, from Eq. (6.52a), $C_2 = 1$ and $G_2 = 0.05$ while, from Eq. (6.52b),

$$C_3 = \frac{1}{1 - (b_1/2)^2} = 1.002 \text{ F} \quad \text{and} \quad G_3 = \frac{(b_1/2)}{1 - (b_1/2)^2} = 0.051 \text{ S}.$$

In Eq. (6.50), $N(s) = Hs$, so that Eq. (6.53) implies

$$-y_{21A} = \frac{Hs}{s + b_1/2}$$

CONTROLLED-SOURCE REALIZATIONS

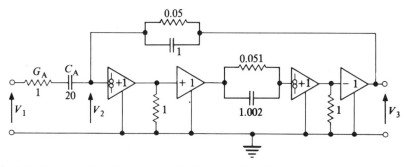

Fig. 6.17. Bandpass filter. Element values in siemens and farads

This suggests the series RC arrangement of Table 3.2 with

$$G_A = H = 1 \text{ S} \quad \text{and} \quad C_A = \frac{2H}{b_1} = 20 \text{ F}$$

The final diagram is shown in Fig. 6.17. The capacitance spread can be alleviated for integrated-circuit realization by choosing $C_A = 1$, $G_A = b_1/2 = H$ at the expense of extra amplification.

6.3.2 Frequency-emphasizing networks

The so-called frequency-emphasizing network (FEN) method[13] is based on the division of the second-order section into a cascade arrangement as in the general Fig. 6.15 with network A as a low-Q asymptotic approximation block and network B comprising an active-correcting block (see compounded responses of Fig. 6.18). The main advantage is the possibility afforded of standardizing building-block topology over a wide Q range.

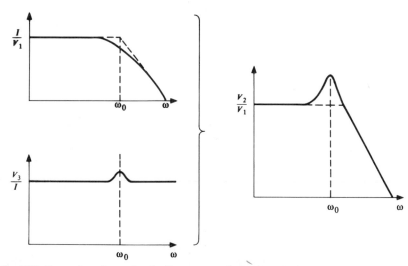

Fig. 6.18. Formation of second-order filter response in FEN synthesis

For medium Q, network A is a passive three-terminal RC network, synthesized to obtain the asymptotic characteristic of the desired function. Network B is a universal correcting network used to provide the desired characteristic by pole-zero cancellation. Hence, to realize the biquadratic function

$$T(s) = H\left(\frac{s^2 + a_1 s + a_0}{s^2 + b'_1 s + b_0}\right) \tag{6.54}$$

the overall function is decomposed so as to be recognizable in the form of Eq. (6.45) as

$$T(s) = -y_{21A} \cdot Z_{21B} = H_A \left(\frac{s^2 + a_1 s + a_0}{s^2 + b_1 s + b_0}\right) H_B \left(\frac{s^2 + b_1 s + b_0}{s^2 + b'_1 s + b_0}\right) \tag{6.55}$$

where $H = H_A H_B$ and coefficients a_i are associated with network A. The transfer admittance y_{21A} can be realized exactly by a passive RC network for non-positive-real zeros and pole $Q \leqslant 0.5$, some suitable structures being given in Table 3.2.

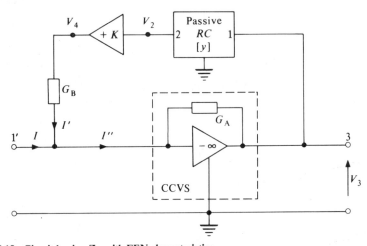

Fig. 6.19. Circuit having Z_{21} with FEN characteristics

An example of an FEN (network B of Fig. 6.15) is shown in Fig. 6.19. Note the presence of the CCVS (Sec. 5.3.4) which ensures zero input impedance, a requirement explained in the previous section. Simple analysis yields the following set of equations:

$$I'' = I + I' \tag{6.56a}$$

transfer characteristic of CCVS, $V_3 = -\left(\dfrac{I''}{G_A}\right)$ \hfill (6.56b)

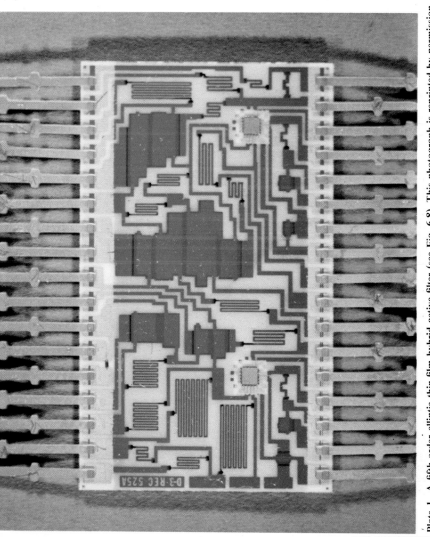

Plate 1. A fifth-order elliptic thin-film hybrid active filter (see Fig. 6.8). This photograph is reprinted by permission from the *Bell Systems Technical Journal* March 1975, © 1975, American Telegraph & Telephone Company

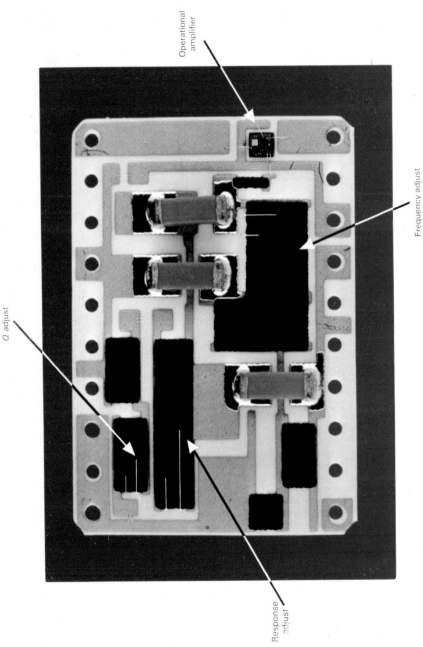

Plate 2. A second-order lowpass/bandpass thick-film hybrid active filter (see Fig. 6.22) illustrating laser trims. By courtesy of Welwyn Electric Limited and acknowledgement to the Admiralty Underwater Weapons Establishment

transfer ratio of RC network, $V_2/V_3 = \dfrac{-y_{21}}{y_{22}}$ (6.56c)

$$V_4 = KV_2 \qquad (6.56d)$$

input impedance of CCVS $= 0$, $\quad I' = G_B V_4$ (6.56e)

from which $\quad Z_{21B} = \dfrac{V_3}{I} = -\left\{\dfrac{1}{G_A - KG_B(y_{21}/y_{22})}\right\}$ (6.57)

Using a general second-order passive network for which

$$\dfrac{-y_{21}}{y_{22}} = \dfrac{s^2 + a_1' s + a_0'}{s^2 + b_1 s + b_0}$$

Eq. (6.57) becomes

$$Z_{21B} = \dfrac{-(s^2 + b_1 s + b_0)\left(\dfrac{1}{G_A + KG_B}\right)}{s^2 + \left(\dfrac{G_A b_1 + KG_B a_1'}{G_A + KG_B}\right) s + \left(\dfrac{b_0 G_A + a_0' KG_B}{G_A + KG_B}\right)} \qquad (6.58)$$

If $a_0' = b_0$, this satisfies the requirements of Eq. (6.55), where

$$b_1' = \dfrac{G_A b_1 + KG_B a_1'}{G_A + KG_B} = b_1 \left(\dfrac{1 + K \cdot \dfrac{a_1'}{b_1} \cdot \dfrac{G_B}{G_A}}{1 + K \dfrac{G_B}{G_A}}\right)$$

and

$$\dfrac{Q_P}{Q_Z} = \dfrac{1 + K \dfrac{G_B}{G_A}}{1 + K \cdot \dfrac{G_B}{G_A} \cdot \dfrac{a_1'}{b_1}}$$

Thus, the system Q factor is only magnified if $a_1' \ll b_1$, emphasizing the need for a symmetrical twin-T giving

$$\dfrac{Q_P}{Q_Z} = \left(1 + K \dfrac{G_B}{G_A}\right) \qquad (6.59)$$

In practice, care must be exercised with the FEN so as to avoid parasitic oscillations. As well as stabilizing the amplifiers individually, it is necessary[13] to select active devices with differing unity-gain bandwidths (see Sec. 9.2), the wider being used for the non-inverting amplifier. Q factors as high as $50 \to 500$ can be obtained by cascading a suitable FEN with a single-source modification of the Sallen and Key networks.

Example 6.5 Design a second-order bandpass filter having a Q factor of 40 and a midband gain of 40 dB.

SOLUTION The frequency-normalized function is

$$T(s) = \frac{Hs}{s^2 + 0.025s + 1}$$

A midband gain of 40 dB $= 100 = H/0.025$. Thus $H = 2.5$. The required form for network A is shown as Circuit No. 7 in Table 3.2. Use of a balanced twin-T in network B sets $a_1' = 0$, $a_0' = b_0 = 1$, and $b_1 = 4$, resulting in the following set of general design equations:

$$b_0 = 1 = \frac{G_1 G_2}{C_1 C_2} \tag{6.60a}$$

$$b_1 = 4 = \frac{G_1 + G_2}{C_1} + \frac{G_2}{C_2} \tag{6.60b}$$

$$b_1' = 0.025 = \frac{4 G_A}{G_A + K G_B} \tag{6.60c}$$

$$H = 2.5 = \left(\frac{G_2 G_1}{C_1}\right)\left(\frac{1}{G_A + K G_B}\right) \tag{6.60d}$$

There are seven unknowns and only four equations, providing the design engineer with a fair degree of choice. Suppose a voltage follower is used and that, for ease of fabrication, the capacitors are made equal with $C_1 = C_2 = 2$ F and $K = 1$. Equations (6.60) may then be solved to yield

$$G_A = 1/200, \quad G_B = 159/200, \quad G_1 = 1.18, \quad \text{and} \quad G_2 = 3.41 \text{ S}$$

The above example shows how a passive Q of 0.25 can be increased to 40 by use of the FEN network. Using the same FEN circuit, a wide range of transfer functions can be designed by simply selecting the appropriate form for network A from Table 3.2.

6.3.3 Combined DVCCS/DVCVS realizations

Although the DVCCS (Sec. 5.3.5) can form a set of generating elements sufficient to realize all linear finite networks, some of the resulting equivalents

CONTROLLED-SOURCE REALIZATIONS

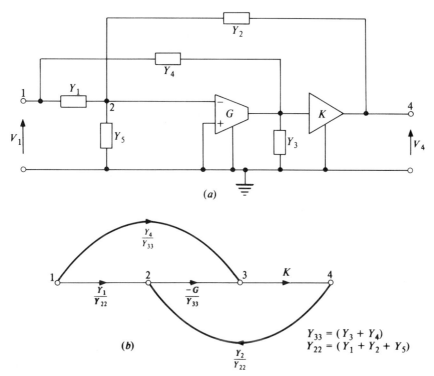

Fig. 6.20. DVCCS/DVCVS system for biquadratic synthesis. (*a*) General biquadratic structure; (*b*) signal-flow graph of (*a*)

are unduly complicated. For practical purposes, therefore, it is convenient to use the single-ended-input mode with an added VCVS and perform synthesis with the resulting mixed-source element.

A structure[14] for the realization of biquadratic functions is shown in Fig. 6.20(*a*). Analysis may be performed with the aid of the signal-flow graph in Fig. 6.20(*b*), from which

$$\frac{V_4}{V_1} = \frac{K\{Y_4 - GY_1/(Y_1 + Y_2 + Y_5)\}}{\{(Y_3 + Y_4) + KY_2G/(Y_1 + Y_2 + Y_5)\}} \quad (6.61)$$

Selection of a divisor polynomial $Q(s) = (s + \sigma)$, $\sigma > 0$, allows a difference-RC decomposition of the numerator and a summation decomposition of the denominator so that

$$T = \frac{Y_{A(RC)} - Y_{B(RC)}}{Z_{A(RL)} + Z_{B(RC)}} \quad (6.62)$$

where $Y_{A(RC)}$, $Y_{B(RC)}$ are RC driving-point admittances and $Z_{A(RL)}$, $Z_{B(RC)}$ are RL and RC driving-point impedances, respectively. Comparing Eqs. (6.61) and (6.62), the driving-point functions can be allocated and synthesized. A

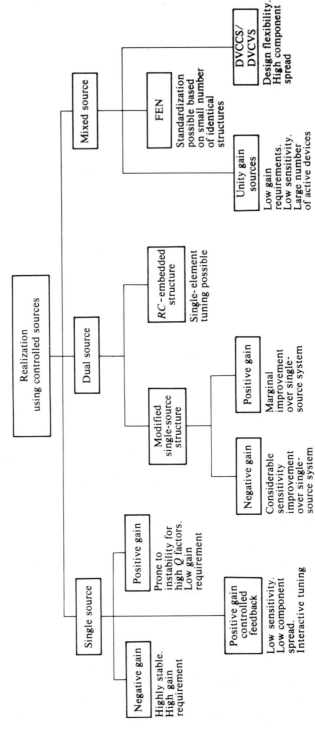

Fig. 6.21. Perspective of techniques for controlled-source realization

suitable realization[14] results if Y_2 consists of a conductance G_2 and the remaining admittances are parallel combinations of capacitance and conductance such that C_1 and G_1 refer to Y_1. When second-order lowpass, bandpass, and highpass functions with no finite zeros are to be realized, this arrangement will not necessarily be canonic in terms of the number of capacitors. However, by suitable choice of design parameters, a two-capacitor solution can be obtained. For example, a lowpass filter has $a_1 = a_2 = 0$, rendering C_4 automatically zero.[14] If $KG = a_0/\sigma$, then C_5 is also zero.

6.4 Optimum design and sensitivity

The methods discussed in this chapter are summarized in Fig. 6.21, which constitutes a guide to the initial choice of design. A comprehensive comparison of the various techniques is afforded by consideration of a bandpass characteristic formed by rewriting Eq. (4.6) as

$$T(s) = \frac{Hs}{s^2 + (\omega_0/Q)s + \omega_0^2} \tag{6.63}$$

The results, presented in Table 6.5, provide information on Q and ω_0 sensitivities together with component spread. To afford comparison with the realizations of later chapters, sensitivity entries in Table 6.5 are listed with respect to the OA open-loop gain. Sensitivities with respect to K (computed from Q and ω_0 expressions) may be obtained by use of the conversion $S_K^Q = S_A^Q/S_A^K$ (from identity 3 of Table 4.2) and the following relationships:

(a) for a non-inverting source of gain K,
$$S_A^K = K/A \quad \text{if } A \gg K;$$
(b) for an inverting source of gain $-K$,
$$S_A^K = (1-K)/A \quad \text{if } A \gg 1,$$

The parameter choice in the design notes of Table 6.5 is that which is most convenient in each case. For the single-source positive-gain system, the choice presented has the advantages of a simple relationship between element values, low component spread, and a low value of VCVS gain. The major disadvantage is its relatively high Q sensitivity to passive elements. This can be reduced by, for example, relaxing the equal-capacitor constraint, an optimum ratio of $C_1/C_2 = 3$ yielding $|S_{RC}^Q|_{max} = \sqrt{3}Q$. Lowpass positive-gain realizations require only four elements and can be designed with passive Q sensitivities independent of Q. The single-source negative-gain arrangement affords improvement in this respect, yielding[15] passive sensitivities several times less than for the equivalent passive LCR resonator. However, this is attended by the disadvantages of high active sensitivities and a high-gain requirement.

ACTIVE FILTERS FOR COMMUNICATIONS AND INSTRUMENTATION

Table 6.5 Performance comparison for second-

Network structure		Performance	
		Q	ω_0
Single source	Positive gain (Fig. 6.2)	$\dfrac{\sqrt{\dfrac{G_1+G_2}{G_3}}}{\sqrt{\dfrac{C_2}{C_1}}\left[1+(1-K)\dfrac{G_2}{G_3}+\dfrac{G_1}{G_3}\right]+\sqrt{\dfrac{C_1}{C_2}}}$	$\sqrt{\dfrac{G_3(G_1+G_2)}{C_1C_2}}$
	Positive gain controlled feedback (Fig. 6.22)	$\dfrac{1}{2}\left(1+\dfrac{G_5}{G_4}\right)\sqrt{\dfrac{G_1}{G_2}\left(\dfrac{C_3}{C_1}-1\right)}$	$\dfrac{1}{C_1}\sqrt{G_1G_2\left(\dfrac{C_3}{C_1}-1\right)}$
	Negative gain (Table 6.1, Cct. No. 4)	$\dfrac{\sqrt{(1-K)\dfrac{G_1}{G_2}}}{\sqrt{\dfrac{C_2}{C_1}}\left(1+\dfrac{G_1}{G_2}\right)+\sqrt{\dfrac{C_1}{C_2}}}$	$\sqrt{\dfrac{G_1G_2}{C_1C_2(1-K)}}$
Dual Source	Positive gain (Fig. 6.11)	$\dfrac{\sqrt{\dfrac{G_1+G_2}{G_3}}}{\sqrt{\dfrac{C_1}{C_2}}+\sqrt{\dfrac{C_2}{C_1}}\left[\dfrac{G_1}{G_3}+\dfrac{G_2}{G_3}(1-K_1K_2)\right]}$	$\sqrt{\dfrac{G_3(G_1+G_2)}{C_1C_2}}$
	Negative gain (Table 6.2 Cct. No. 4)	$\dfrac{\sqrt{(1-K_1K_2)\dfrac{G_1}{G_2}}}{\dfrac{G_1}{G_2}\sqrt{\dfrac{C_2}{C_1}}+\sqrt{\dfrac{C_1}{C_2}}}$	$\sqrt{\dfrac{G_1G_2}{C_1C_2(1-K_1K_2)}}$
RC embedded	Summation-input twin-T (Table 6.4)	$\dfrac{1+\dfrac{G_B}{G_A}}{4\left[1+\dfrac{G_B}{G_A}(1-K_1K_2)\right]}$	$\dfrac{G}{C}$
	Difference-input bridged-T (Table 6.4)	$\dfrac{1}{2+n-K_2\left(1+\dfrac{G_A}{G_B}\right)}$	$\dfrac{G}{C}$
Mixed source	Unity gain (Fig. 6.16)	$\dfrac{\sqrt{\left(1+\dfrac{G_1G_4}{G_2G_3}\right)\left(\dfrac{1}{K_1K_2K_3K_4}\right)}}{\sqrt{\dfrac{C_3}{C_2}}\sqrt{\dfrac{G_2}{G_3}}+\sqrt{\dfrac{C_2}{C_3}}\sqrt{\dfrac{G_3}{G_2}}}$	$\sqrt{\dfrac{G_2G_3+\dfrac{G_1G_4}{K_1K_2K_3K_4}}{C_2C_3}}$
	FEN (Fig. 6.19)	$\dfrac{\left(1+\dfrac{KG_B}{G_A}\right)\sqrt{\dfrac{G_1}{G_2}}}{\sqrt{\dfrac{C_2}{C_1}}\left(1+\dfrac{G_1}{G_2}\right)+\sqrt{\dfrac{C_1}{C_2}}}$	$\sqrt{\dfrac{G_1G_2}{C_1C_2}}$
	DVCCS/DVCVS (Fig. 6.20)	$\dfrac{\sqrt{\dfrac{(G_3+G_4)}{G_2}\left(1+\dfrac{G_1}{G_2}+\dfrac{G_5}{G_2}\right)+K\dfrac{G}{G_2}}}{\sqrt{\dfrac{C_3}{C_5}}\left(1+\dfrac{G_1}{G_2}+\dfrac{G_5}{G_2}\right)+\sqrt{\dfrac{C_5}{C_3}}\left(\dfrac{G_3+G_4}{G_2}\right)}$	$\sqrt{\dfrac{(G_3+G_4)(G_1+G_2+G_5)+KGG_2}{C_3C_5}}$

CONTROLLED-SOURCE REALIZATIONS

...rder controlled-source RC bandpass filter realizations

...arameters for bandpass transfer function

S_A^Q	$S_A^{\omega_0}$	$\|S_{RC}^Q\|_{max}$	$\|S_{RC}^{\omega_0}\|_{max}$	Maximum component spread	Design conditions
$\dfrac{9Q}{A}$	0	$\approx 2Q$	$\dfrac{1}{2}$	2	$K = (3 - 1/Q) = H$ $G_1 = G_2 = 2G_3 \quad C_1 = C_2$
$\dfrac{4Q}{A}$	0	$\dfrac{Q}{2}$	$\dfrac{3}{8}$	$2Q$	$K = 1 \quad G_1 = G_2 = G_3/2 \quad G_5 = (2Q - 1)G_4$ $C_1 = C_2 = C_3/2$
$\dfrac{-9Q^2}{2A}$	$\dfrac{9Q^2}{2A}$	$\dfrac{1}{6}$	$\dfrac{1}{2}$	$9Q^2$	$K = 1 - 9Q^2 \quad G_1 = G_2 \quad C_1 = C_2$ $H = K/\sqrt{1 - K}$
$\dfrac{4Q}{A}$	0	$\dfrac{3Q}{2}$	$\dfrac{1}{2}$	2	$K_1 K_2 = \left(\dfrac{5}{2} - \dfrac{1}{Q}\right) = H \quad G_1 = G_2 = 2G_3$ $C_1 = C_2$
$\dfrac{-2Q^2}{A}$	$\dfrac{2Q^2}{A}$	0	$\dfrac{1}{2}$	$4Q$	$K_1 K_2 = 1 - 4Q^2 \quad G_1 = G_2 \quad C_1 = C_2$ $H = K_1 K_2/\sqrt{1 - K_1 K_2}$
$\dfrac{4Q}{A}$	0	$\dfrac{Q}{2}$	$\dfrac{3}{8}$	$4Q$	$K_1 = K_2 = 1, \quad G_1 = G_2 = G_3/2$ $C_1 = C_2 = C_3/2, \quad G_B/G_A = 4Q - 1, \quad H = 1/Q$ $Q \gg 1$
$\dfrac{2Q}{A}$	0	Q	$\dfrac{1}{2}$	Q	$K_2 = 1, \quad A_1 = A_2 = A, \quad G_1 = G_2/n, \quad C_1 = C_2/n$ $n = 1/Q \quad G_A = G_B \quad H = K_2/q = 1 \quad Q \gg 1$
$\dfrac{-1}{2A_2} \quad \dfrac{1}{A_4}$	$\dfrac{-1}{2A_2} \quad \dfrac{1}{A_4}$	$\dfrac{1}{2}$	$\dfrac{1}{2}$	$2Q$	$G_1 = 2QG_2 = \dfrac{4Q^2 - 1}{2Q} G_3 = G_4 = G_A/H$ $C_2 = \dfrac{4Q^2 - 1}{4Q^2} \cdot C_3 = C_A/2QH$ $K_1 = K_2 = K_3 = K_4 = 1$
$\left(4\dfrac{G_A}{G_B}\right)\dfrac{Q}{A}$	0	≈ 1	$\dfrac{1}{2}$	$4Q$	$K = 1 \quad C_1 = C_2 = k$ $G_1 = k^2/G_2 = 4QH(2 \pm \sqrt{2})G_A$ $= \dfrac{KH(2 \pm \sqrt{2})}{(1 - 1/4Q)} G_B, \quad Q \gg 1$
$\dfrac{HQ^2}{A}$	$\dfrac{1}{2}$	$\dfrac{1}{2}$	$\dfrac{1}{2}$	$4Q^2$	$K = 2QH, \quad C_1 = C_4 = G_3 = 0, \quad C_3 = \dfrac{(4Q^2 - 1)}{4Q^2(2Q + H)} \cdot C_5$ $G_1 = \dfrac{2QH}{4Q^2 - 1} \quad G_2 = \dfrac{4Q^2 H}{4Q^2 - 1} \quad G_4 = 2QHG_5$ $= \dfrac{(4Q^2 - 1)(H + 2Q)}{4Q^2 H^2} G$

Whereas the dual-source positive-gain system offers only marginal advantages in passive Q sensitivity compared to its single-source counterpart, significant improvement is obtained in the negative-gain case. The individual source gains are reduced and, more importantly, all passive Q sensitivities are identically zero. However, the negative-gain systems both have $S_A^{\omega_0}$ proportional to Q^2 while the positive-gain designs have zero which, by Eq. (4.39), could be an important performance factor.

Of the RC-embedded structures, the summed-input twin-T arrangement of Table 6.4 offers tuning advantages over the difference-input bridged-T circuit, as discussed in Sec. 6.2.2. Furthermore, the lower passive Q and ω_0 sensitivities are more practically significant than the higher active Q sensitivity (see Chapter 4). It should be noted at this junction that the ω_0 expressions for each network are based on the particular component relationships described in Sec. 6.2.2. Evaluation of $|S_{RC}^{\omega_0}|_{max}$, on the other hand, must refer to the exact expressions for ω_0 and the results presented in Table 6.5 cannot be obtained from the simplified formulae.

The mixed-source unity-gain system combines relatively low sensitivities with low gain requirement but assumes ideal current sources, exhibits high component spread (or low gain) and can be unstable in practice. The second arrangement in this category, the FEN system, has the advantages of being suitable for large-scale production as well as having independent tuning facilities for poles and zeros. Though relatively complex, the network of Fig. 6.19 is the same for all systems having similar Q requirements. Calculations for the sensitivities listed in Table 6.5 ignore the imperfections of the constituent operational amplifier in the CCVS. The maximum component spread is a representative figure, as may be seen from Ex. 6.5, being tabulated as the ratio of (gain-defining components) G_B to G_A. The overall spread is often several times greater in specific cases, according to the choice of network A. Thus, in the example cited, the maximum ratio is $(G_2/G_A) \approx 17Q$ rather than $4Q$.

The DVCCS/DVCVS realization, though suffering from high component spread, combines design flexibility with low-sensitivity parameters, with the exception of S_A^Q which is proportional to Q^2. It is important to observe that the gain G of the DVCCS must be treated as an additional network conductance in so far as both denormalization and sensitivity are concerned. Hence

$$\sum_i S_{G_i}^Q = 0 \quad \text{only if } S_G^Q \text{ is included}$$

and

$$\sum_i S_{G_i}^{\omega_0} = -1 \quad \text{only if } S_G^{\omega_0} \text{ is included.}$$

Single-amplifier realizations of comparable bandpass characteristics are obtainable by rotating the twin-T network configuration of Fig. 6.7,

CONTROLLED-SOURCE REALIZATIONS

interchanging input and earth terminals as in Fig. 6.22. On account of the previously mentioned tuning interaction, it is necessary to trim three resistors for precise Q and centre frequency. This is illustrated in Plate 2 for a popular lowpass filter adaptation manufactured in thick-film hybrid form.

The relative merits of the various designs considered in this chapter are strongly dependent upon the form of practical implementation. For instance, the advantages stated for the more complex configurations incorporating RC twin-T networks are reliant on construction in integrated technology. Discrete versions of the twin-T prove too expensive to implement and component

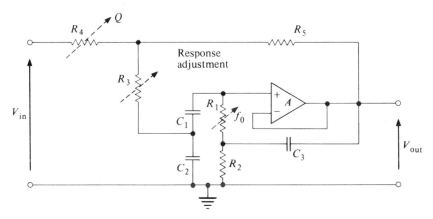

Fig. 6.22. Controlled positive-feedback Lim bandpass filter

mismatch renders tuning difficult. On the other hand, as will be seen in Chapter 9, thin and thick-film realizations are readily trimmed to precision and are amenable to mass-production methods which take full advantage of the flexibility offered by the greater complexity.

References

1. L. P. Huelsman (Ed.), *Active Filters: Lumped, Distributed, Integrated, Digital, and Parametric*, McGraw-Hill, New York, 1970, Chap. 2.
2. R. P. Sallen and E. L. Key, 'A practical method of designing RC active filters', *IRE Trans.*, **CT-2**, 74–85, 1955; also reprinted in *Active Inductorless Filters* (S. K. Mitra, ed.), IEEE Press, New York, 1971.
3. J. T. Lim, 'Improvements in or relating to active filter networks', British Patent Application 9657, 16th April 1971.
4. A. S. Sedra, 'A class of stable active filters using unity-gain voltage followers', *IEEE Journal* **SC-7**, 311–315, 1972.
5. R. M. Inigo, 'Active filter realization using finite-gain voltage amplifiers', *IEEE Trans.*, **CT-17**, 445–448, 1970; also reprinted in *Active Inductorless Filters* (S. K. Mitra, Ed.), IEEE Press, New York, 1971.

6. C. Daryanani, 'Active filters—what's available in the industry', in *Proceedings of the International Symposium on Circuit Theory, Toronto*, IEEE, New York, 1973, pp. 398–401.
7. S. C. Dutta Roy and K. K. Malik, 'Active RC realization of a third-order low-pass Butterworth characteristic', *Electron. Letters*, **8**, 630–631, 1972.
8. F. W. Stephenson, 'RC active filters for microelectronic realization', *Microelectronics*, **6** (3), 7–10, 1975.
9. G. Wilson, Y. Bedri, and P. Bowron, 'RC-active networks with reduced sensitivity to amplifier gain-bandwidth product', *IEEE Trans.*, **CAS-21** (5), 618–626, 1974.
10. A. M. Soliman, 'A novel canonic active RC bandpass network with reduced sensitivity to amplifier gain bandwidth product', in *Proceedings of ISCAS Munich, 1976*, IEEE Press, New York, 1976, pp. 493–496.
11. A. M. Soliman, 'New active RC configuration for realizing a medium-selectivity notch filter', *Electron. Letters*, **8** (21), 522–524, 1972.
12. E. H. C. Cheng and K. A. Stromsoe, 'Synthesis of integrable high-Q filters using grounded unity-gain amplifiers', *Int. J. Electron.*, **33** (6), 665–676, 1972.
13. G. S. Moschytz, 'FEN filter design using tantalum and silicon integrated circuits', *Proc. IEEE*, **58** (4), 550–566, 1970.
14. M. Bialko, W. Sienko, and R. W. Newcomb, 'Active synthesis using the DVCCS/DVCVS', *Int. J. Circuit Theory Applics*, **2**, 23–28, 1974.
15. P. R. Geffe, 'RC-amplifier resonators for active filters', *IEEE Trans.*, **CT-15**, 415–419, 1968.

7. Realization using single operational amplifiers

The operational amplifier has been introduced in Chapter 5 and used to simulate various active elements. Synthesis methods based on the controlled sources thereby arising have subsequently been developed in Chapter 6. The intention now is to consider the direct application of the operational amplifier itself as the active component in RC filters without the intermediary of other active element concepts. The special features arising from the high-gain capabilities of this amplifier merit a separate treatment. Such circuits are a generalization and extension of analogue computer techniques, exploiting the RC synthesis procedures of Chapter 3 and the differential-input facilities of the operational amplifier to full advantage.

7.1 Single-feedback circuits

The use of a passive RC section as the single negative-feedback path of an operational amplifier is one of the earliest and commonest methods of realizing

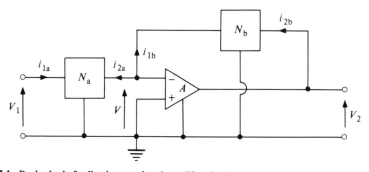

Fig. 7.1. Basic single-feedback operational amplifier circuit

voltage transfer functions.[1] In the basic configuration of Fig. 7.1, the non-inverting input to the amplifier is earthed and passive three-terminal RC networks N_a and N_b connected in the feedforward and feedback links, respectively, to the inverting input. If the two-port admittance parameters of the passive sections are given by $[y]_a$ and $[y]_b$ (see Chapter 3) then the virtual-earth property at the single input node implies that the total current

$$i_{2a} + i_{1b} = 0$$

or

$$y_{21a}V_1 + y_{22a}V + y_{11b}V + y_{12b}V_2 = 0$$

Since $V = -V_2/A$, the open-circuit voltage transfer ratio may be written as

$$\frac{V_2}{V_1} = \frac{-y_{21a}}{y_{12b} - (1/A)(y_{22a} + y_{11b})} \qquad (7.1)$$

For the ideal operational amplifier, $A \to \infty$, so that

$$V_2/V_1 = -y_{21a}/y_{12b} \qquad (7.2)$$

Networks N_a and N_b have their natural frequencies on the negative-real axis of the complex-frequency plane, but, if they are equal, cancellation ensures that they will not affect the voltage transfer function of the overall network. If the function to be realized is

$$V_2/V_1 = N(s)/D(s)$$

then identify

$$y_{21a} = -N(s)/Q(s) \quad \text{and} \quad y_{12b} = -D(s)/Q(s) \qquad (7.3)$$

where $Q(s)$ is an auxiliary polynomial containing the natural frequencies. Thus, the poles of the transfer function are determined only by the zeros of transfer admittance y_{12b}. As transformerless passive RC networks can have transmission zeros anywhere in the complex plane except on the positive-real axis, then realization of any stable transfer function is enabled with the exception of the odd-order allpass.

The realization problem thus reduces to the specific synthesis of the constituent passive networks using the techniques of Chapter 3. Should $N(s)$ and $D(s)$ contain complex roots, then it is necessary to invoke parallel-ladder or other non-simple methods.[2] However, for first- and second-order admittance functions, this may be circumvented by resort to pre-analysed[3,4] RC two-port transfer admittances as presented in Table 3.2.

Example 7.1 Realize a second-order bandpass transfer function giving a response Q of 10 and centre-band gain of 20 dB. Denormalize to a centre frequency of 10 kHz, setting all components to values suitable for fabrication as thin-film components.

REALIZATION USING SINGLE OPERATIONAL AMPLIFIERS

SOLUTION The frequency-normalized transfer function is of the form

$$\frac{V_2}{V_1} = \frac{-Hs}{s^2 + bs + 1}$$

where

$$b = \frac{1}{10} \quad \text{and} \quad H = b \text{ antilog}_{10} 1 = 1.$$

With $Q(s) = s + \sigma$, then

$$y_{21a} = \frac{-Hs}{s + \sigma}$$

suggesting the series circuit comprising R and C in Fig. 7.2, for which

$$y_{21} = \frac{-(1/R)s}{s + 1/RC}.$$

Therefore

$$R = 1/H = 1\ \Omega \quad \text{and} \quad c = H/\sigma$$

Also, assuming that $RC = \dfrac{C_2 R_1 R_2}{(R_1 + R_2)}$,

$$-y_{12b} = \frac{s^2 + bs + 1}{s + \sigma} = s + (b - \sigma) + \frac{(1 + \sigma^2 - \sigma b)}{(s + \sigma)}$$

suggesting a paralleled capacitance, resistance, and T-network, i.e., the bridged-T in Table 3.2 comprising R_1, R_2, R_3, C_1, and C_2. The resulting network is clearly simplified by choosing $\sigma = b = 0.1$, implying the absence of R_3. Then, obviously, $C_1 = 1$ F, while $C = 10$ F.

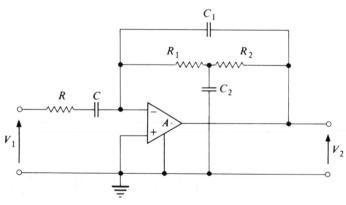

Fig. 7.2. Single-feedback bandpass filter

The T-network has

$$y_{12} = \frac{\dfrac{-1}{C_2 R_1 R_2}}{s + \dfrac{1}{C_2}\left(\dfrac{1}{R_1} + \dfrac{1}{R_2}\right)}$$

Minimum resistive spread requires $R_1 = R_2 = \sigma/2 = 0.05 \,\Omega$, so that

$$C_2 = 4/\sigma^2 = 400 \text{ F}$$

To denormalize, note that

$$-y_{12b} = \frac{s^2 C_1 + s\dfrac{(R_1 + R_2)}{R_1 R_2} \cdot \dfrac{C_1}{C_2} + \dfrac{1}{R_1 R_2 C_2}}{s + \dfrac{R_1 + R_2}{C_2 R_1 R_2}}$$

and the band-centre frequency is given by equating the real part of the transfer function denominator to zero as

$$\omega_0 = \frac{1}{\sqrt{R_1 R_2 C_1 C_2}} = \frac{1}{20 R_1 C_1}$$

or

$$R_1 C_1 = \frac{1}{40\pi \times 10^4} = 0.796 \times 10^{-6}$$

Taking $C_1 = 100$ pF as an easily integrable value gives (see Appendix 1)

$R_1 = R_2 = 7.96$ kΩ, $R = 159.2$ kΩ, $C_2 = 0.04\,\mu\text{F}$ and $C = 1000$ pF.

7.1.1 High-Q realizations

Applications of bandpass filters often require a high Q, i.e., a small coefficient b, so that the poles of the transfer function lie close to the $j\omega$ axis. The bridged-T network is not adequate for this purpose as the expressions in Ex. 7.1 indicate impractical component ratios. Consider again the transfer admittance of N_b:

$$-y_{12b} = \frac{s^2}{s + \sigma} + \frac{1 - \sigma b}{s + \sigma} + b \tag{7.4}$$

REALIZATION USING SINGLE OPERATIONAL AMPLIFIERS

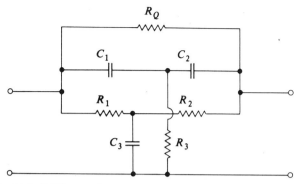

Fig. 7.3. Bridged twin-T RC network

suggesting paralleled RC highpass and lowpass T networks and a resistance (see Fig. 7.3). The highpass T network has

$$-y_{12H} = \frac{\dfrac{C_1 C_2}{C_1 + C_2} s^2}{s + \dfrac{1}{R_3(C_1 + C_2)}} \tag{7.5}$$

while the lowpass T network has

$$-y_{12L} = \frac{\dfrac{1}{R_1 R_2 C_3}}{s + \dfrac{1}{C_3}\left(\dfrac{1}{R_1} + \dfrac{1}{R_2}\right)} \tag{7.6}$$

and, on account of the factor of a half emerging in the symmetrical case,

$$R_Q = 2/b \tag{7.7}$$

As in Table 3.2, the component values therefore become

$$R = \frac{1}{H}, \quad R_1 = R_2 = \frac{\sigma}{2(1-\sigma b)}, \quad R_3 = \frac{1}{4\sigma}$$

$$C = H/\sigma, \quad C_1 = C_2 = 2, \quad C_3 = \frac{4(1-\sigma b)}{\sigma^2}$$

and, for the above example, this time taking $\sigma = 1$,

$$R = 1\,\Omega, \quad R_1 = R_2 = 0.556\,\Omega, \quad R_3 = 0.25\,\Omega, \quad R_Q = 20\,\Omega$$
$$C = 1\,\text{F}, \quad C_1 = C_2 = 2\,\text{F}, \quad C_3 = 3.6\,\text{F}.$$

A reduction in spread of capacitance is evident since the component ratio is a first rather than a second-order function of the required selectivity. Notice that the realization has assumed that

$$\sigma = \frac{1}{R_3(C_1 + C_2)} = \frac{1}{C_3}\left(\frac{1}{R_1} + \frac{1}{R_2}\right) \qquad (7.8)$$

i.e., the well-known condition[5] for reduction of the third-order twin-T transfer function to the second-order null form of Eq. (3.34) by real pole-zero cancellation. Bridging resistor R_Q serves to shift the zero into the left half-plane for a finite Q. This would, of course, be achieved by unbalancing the twin-T itself and this practice is inconvenient and attended by severe component interactions. On the other hand, Eq. (7.7) indicates that R_Q effects complete and accurate control of selectivity within the limits of thin-film resistor fabrication bounds. Alternatively, the designer may again prefer to apply standard tables of admittance functions.[3,4]

Other types of response function may be realized in like manner. With resistors and capacitors interchanged in N_a of the lowpass network, a highpass response results,[4] while use of a twin-T for N_a allows a notch function.[3]

7.2 Multiple-loop feedback circuits

Another method employing a single amplifier in its operational mode is to interconnect a number of two-terminal passive elements in the negative feedback path, so forming the multiple-loop feedback (MLFB) filter. A wide range of configurations may be considered to fall within this category. The network[6] in Fig. 7.4 when analysed with the aid of Kirchhoff's current law has,

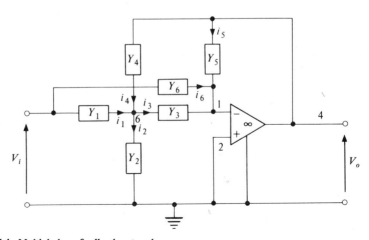

Fig. 7.4. Multiple-loop feedback network

REALIZATION USING SINGLE OPERATIONAL AMPLIFIERS

at nodes 6 and 1,

$$i_1 + i_4 - i_2 - i_3 = 0 \qquad (7.9a)$$
$$i_3 + i_5 + i_6 = 0 \qquad (7.9b)$$

so that, in terms of voltages and admittances,

$$(V_i - V_6)Y_1 + (V_o - V_6)Y_4 - V_6 Y_2 - (V_6 - V_1)Y_3 = 0 \qquad (7.10a)$$
$$(V_6 - V_1)Y_3 + (V_o - V_1)Y_5 + (V_i - V_1)Y_6 = 0 \qquad (7.10b)$$

The virtual-earth point imposed by the operational amplifier ensures that $V_4 \to 0$. Hence, Eqs. (7.10) yield the voltage transfer ratio:

$$\frac{V_o}{V_i} = -\frac{Y_1 Y_3 + Y_6(Y_1 + Y_2 + Y_3 + Y_4)}{Y_3 Y_4 + Y_5(Y_1 + Y_2 + Y_3 + Y_4)} \qquad (7.11)$$

which, though capable of realizing biquadratic and (with $Y_2 = Y_6 = 0$) all-pole functions,[7] does not yield notch functions except with $Y_2 = 0$ and either double-element replacements[8] for the remaining admittances or by cascading[9] with a passive zero-forming section. The most widely used of the multiple-loop arrangements is the double-ladder or Rauch structure of Fig. 7.5, formed by setting $Y_6 = 0$. From Eq. (7.11) the transfer ratio is

$$\frac{V_o}{V_i} = \frac{-Y_1 Y_3}{Y_5(Y_1 + Y_2 + Y_3 + Y_4) + Y_3 Y_4} \qquad (7.12)$$

Table 7.1 shows the network in its low-, high-, and bandpass forms. These are not unique; for instance, to obtain a lowpass function one need only ensure that Y_1 and Y_3 are conductances, Y_5 is a capacitor, and either Y_4 or Y_2 is a

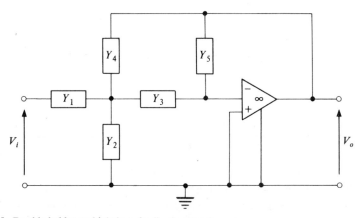

Fig. 7.5. Double-ladder multiple-loop feedback network

ACTIVE FILTERS FOR COMMUNICATIONS AND INSTRUMENTATION

Table 7.1 Multiple-loop *RC* active filters

Filter type	Network	Voltage transfer function
(a) Lowpass		$\dfrac{-G_1 G_3}{s^2 C_2 C_5 + s C_5 (G_1 + G_3 + G_4) + G_3 G_4}$
(b) Highpass		$\dfrac{-s^2 C_1 C_3}{s^2 C_3 C_4 + s G_5 (C_1 + C_3 + C_4) + G_2 G_5}$
(c) Bandpass		$\dfrac{-s G_1 C_3}{s^2 C_3 C_4 + s G_5 (C_3 + C_4) + G_5 (G_1 + G_2)}$

capacitor. Design can be performed by coefficient matching. Consider the lowpass voltage transfer function

$$\frac{V_o}{V_i} = \frac{H}{b_2 s^2 + b_1 s + b_0} \qquad (7.13)$$

By comparing the coefficients for the lowpass network shown in Table 7.1 with those in Eq. (7.13), the following design equations result:

$$G_1 G_3 = H \qquad (7.14a)$$
$$C_2 C_5 = b_2 \qquad (7.14b)$$
$$C_5 (G_1 + G_3 + G_4) = b_1 \qquad (7.14c)$$
$$G_3 G_4 = b_0 \qquad (7.14d)$$

These equations do not yield an equal-capacitor solution but may be solved for equal-resistor pairs, the resulting element values for $b_2 = 1$ and $G_3 = G_4$ being shown in Table 7.2. A second equal-resistor solution is possible if $G_1 = G_3$ while, for $G_1 = G_4$, a realizable network only ensues if $H = b_0$.

REALIZATION USING SINGLE OPERATIONAL AMPLIFIERS

Table 7.2 Element values and component spread for the MLFB structure

Element (Fig. 7.5)	Element values Lowpass	Element values Highpass	Element values Bandpass
Y_1	$G_1 = \dfrac{H}{\sqrt{b_0}}$	$C_1 = H$	$G_1 = H$
Y_2	$C_2 = \dfrac{(2b_0 + H)}{b_1 \sqrt{b_0}}$	$G_2 = \dfrac{b_0(2 + H)}{b_1}$	$G_2 = \dfrac{2b_0}{b_1} - H$
Y_3	$G_3 = \sqrt{b_0}$	$C_3 = 1$	$C_3 = 1$
Y_4	$G_4 = G_3$	$C_4 = C_3$	$C_4 = C_3$
Y_5	$C_5 = \dfrac{b_1 \sqrt{b_0}}{(2b_0 + H)}$	$G_5 = \dfrac{b_1}{(2 + H)}$	$G_5 = \dfrac{b_1}{2}$
G spread	$\dfrac{H}{b_0}$	$\dfrac{b_0(2 + H)^2}{b_1^2}$	$\dfrac{2}{b_1}\left(\dfrac{2b_0}{b_1} - H\right)$
C spread	$\dfrac{(2b_0 + H)^2}{b_1^2 b_0}$	H	1

Example 7.2 Design a lowpass Butterworth filter having a passband gain of 20 dB. It will suffice to consider only the normalized solution with a cut-off at $\omega = 1$ rad/s.

SOLUTION The network form is shown in Table 7.1 and the desired transfer function has $H = 10$, $b_1 = \sqrt{2}$, and $b_0 = 1$. Solving Eq. (7.14) for equal-resistor pairs yields two sets of components: namely,

case 1 (see Table 7.2): $G_1 = 10$, $G_3 = G_4 = 1$, $C_2 = 8.485$, and $C_5 = 0.118$
case 2: $G_1 = G_3 = 3.162$, $G_4 = 0.316$, $C_2 = 4.696$, and $C_5 = 0.213$

where the element values are in siemens and farads. The two solutions exhibit equal conductance spread whereas case 2 yields a lower capacitance ratio and, in general, does so for $H > b_0$.

Retaining the admittance equality $Y_3 = Y_4$ of the lowpass example, the element values for equal-capacitor highpass and bandpass networks are listed in Table 7.2. The conditions for maximum Q in the bandpass arrangement are: $C_3 = C_4$ and $G_1 = G_2 = nG_5$, where n is a real number. This yields $Q_{max} = \sqrt{(n/2)}$ and can be obtained from case 2 in Table 7.2 if $H = (b_0/b_1)$ and thus if $n = (2b_0/b_1^2)$.

It is clear from Eq. (7.12) that systems containing imaginary zeros cannot be realized by this configuration. This disadvantage can be overcome through the utilization[11] of an extended node-introduction synthesis theory.

7.3 Differential-input amplifier circuits

The methods described in Secs. 7.1 and 7.2 have employed a single-ended operational amplifier. By exploiting the dual-input facility of most commercially available integrated operational amplifiers (DIOA's), greater versatility can be obtained without the necessity for introducing additional single-ended stages.

7.3.1 Differential-input single-feedback circuits[12]

The general configuration of Fig. 7.6 enables realization of the same range of transfer functions as the single-feedback system of Sec. 7.1, but unattended by

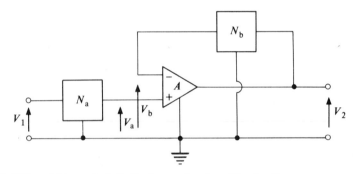

Fig. 7.6. Differential-input single-feedback operational amplifier circuit

overall phase inversion. If the open-circuit voltage transfer functions of passive RC networks N_a and N_b are $T_a(s)$ and $T_b(s)$, respectively, then with the notation indicated, analysis gives

$$V_2 = A(V_a - V_b)$$

or

$$V_2/A = T_a V_1 - T_b V_2$$

leading to

$$\frac{V_2}{V_1} = \frac{T_a}{T_b + 1/A} \qquad (7.15)$$

and, for A infinite, to

$$\frac{V_2}{V_1} = \frac{T_a}{T_b} \qquad (7.16)$$

Like Eq. (7.2), this allows all but positive-real-axis zeros. The poles of T_a and T_b being chosen to cancel, synthesis resolves into the separate realization of the two constituent passive functions either by the procedures of Sec. 7.1 or by reference to tabulations.[3] For the normalized second-order case,

$$T_a(s) = \frac{N(s)}{s^2 + bs + 1} \qquad (7.17)$$

is realizable by a variety of simple ladder networks while

$$T_b(s) = \frac{s^2 + (1/Q)s + 1}{s^2 + bs + 1} \qquad (7.18)$$

may, again, be a bridged-T or bridged twin-T.

Substituting Eqs. (7.17) and (7.18) into Eq. (7.15) for finite A gives

$$\frac{V_2}{V_1} = \frac{1}{1 + 1/A} \cdot \frac{N(s)}{s^2 + (1/Q)(1 + Q_b/A)s + 1} \qquad (7.19)$$

with effective selectivity

$$Q_k = \frac{Q}{1 + Q_b/A} \qquad (7.20)$$

On the other hand, the bridged-balanced twin-T of Fig. 7.3 with $R_1 = R_2 = R$, $R_3 = R/2$, $C_1 = C_2 = C$, $C_3 = 2C$, $R_\varrho = R/a$ leads to

$$\frac{V_2}{V_1} = \frac{N(s)}{s^2 + (4/A + 2a)s + (1 + 2a)} \qquad (7.21)$$

so that

$$\omega = \omega_0 \sqrt{1 + 2a}$$

and

$$Q_k \simeq \frac{\sqrt{1 + 2a}}{2a + 4/A} \simeq \frac{1}{2a} \qquad (7.22)$$

The small resultant shift in centre frequency can readily be compensated by predistortion (Appendix 2) of the twin-T design. In this case, it is necessary that RC network N_a assumes the form shown in Fig. 7.7, otherwise response functions are restricted. The advantage is that the differential-input connection avoids increased sensitivity caused by interaction between the two networks N_a and N_b inherent in Eq. (7.1). The order of sensitivity is, however, still that dictated by choice of N_b.

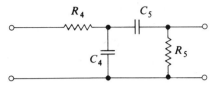

Fig. 7.7. Passive *RC* network N$_a$ for the differential-input single-feedback bandpass filter

7.3.2 General differential-input circuits

The restriction to transfer functions without positive-real poles and zeros may be removed by introducing additional single-ended infinite-gain amplifiers. Alternatively it is more economical to utilize again the differential-input facility afforded by modern integrated operational amplifiers.

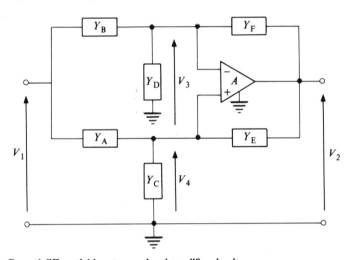

Fig. 7.8. General differential-input operational amplifier circuit

Analysing the *RC*-active configuration[3] of Fig. 7.8 gives

$$Y_B(V_1 - V_3) + Y_F(V_2 - V_3) - Y_D V_3 = 0 \qquad (7.23)$$

$$Y_A(V_1 - V_4) + Y_E(V_2 - V_4) - Y_C V_4 = 0 \qquad (7.24)$$

$$V_2 = A(V_4 - V_3) \qquad (7.25)$$

combining into transfer ratio

$$\frac{V_2}{V_1} = \frac{Y_A(Y_B + Y_D + Y_F) - Y_B(Y_A + Y_C + Y_E)}{Y_F(Y_A + Y_C + Y_E) - Y_E(Y_B + Y_D + Y_F) + (1/A)(Y_A + Y_C + Y_E)(Y_B + Y_D + Y_F)} \qquad (7.26)$$

For convenience, choose

$$Y_A + Y_C + Y_E = Y_B + Y_D + Y_F \qquad (7.27)$$

REALIZATION USING SINGLE OPERATIONAL AMPLIFIERS

and, as $A \to \infty$, then

$$\frac{V_2}{V_1} = \frac{Y_A - Y_B}{Y_F - Y_E} \tag{7.28}$$

This form indicates that the configuration of Fig. 7.8 is capable of realizing any real transfer function, $N(s)/D(s)$. Dividing numerator and denominator by a polynomial $Q(s)$ with simple negative-real roots and of order at least one less than that of $N(s)$ or $D(s)$, the resulting admittances Y_A, Y_B, Y_E, Y_F may be found as passive RC driving-point functions. The remaining one-ports are identified by writing, from Eq. (7.27),

$$Y_C - Y_D = (Y_F - Y_E) - (Y_A - Y_B)$$

$$= \frac{D(s) - N(s)}{Q(s)} \tag{7.29}$$

and associating terms.

Example 7.3 Realize the bandpass voltage transfer function

$$T(s) = \frac{s}{s^2 + 0.1s + 1}$$

using a DIOA.

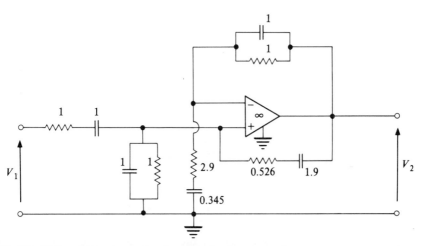

Fig. 7.9. Differential-input operational amplifier bandpass example

SOLUTION Choose $Q(s) = s + 1$, giving

$$\frac{N(s)}{Q(s)} = \frac{s}{s+1} = \frac{s}{s+1} - 0 = Y_A - Y_B$$

$$\frac{D(s)}{Q(s)} = \frac{s^2 + 0.1s + 1}{s+1} = (s+1) - \frac{1.9s}{s+1} = Y_F - Y_E$$

$$\frac{D(s) - N(s)}{Q(s)} = \frac{s^2 - 0.9s + 1}{s+1} = (s+1) - \frac{2.9s}{s+1} = Y_C - Y_D$$

Associating RC admittances gives the circuit in Fig. 7.9.

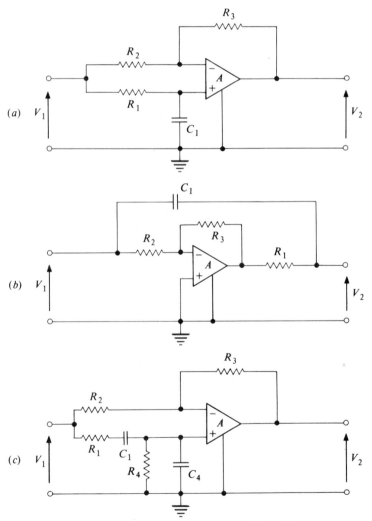

Fig. 7.10. Allpass circuits. (a) Lagging (note that interchange of R_1 and C_1 gives leading circuit); (b) single-input; (c) second-order Wien bridge

REALIZATION USING SINGLE OPERATIONAL AMPLIFIERS

Allpass sections   A useful degeneration of Fig. 7.8 results when $Y_D = Y_E = 0$. Then, if $R_1 = 1/Y_A$, $R_2 = 1/Y_B$, $R_3 = 1/Y_F$ and $sC_1 = Y_C$, the simplified circuit[13,14] of Fig. 7.10(a) (see also Fig. 5.27(a)) has the transfer function

$$\frac{V_2}{V_1} = \frac{1 - sC_1 R_1 (R_3/R_2)}{1 + sC_1 R_1} \qquad (7.30)$$

Interchanging R_1 and C_1 gives the allpass leading transfer function:

$$\frac{V_2}{V_1} = \frac{sCR_1 - (R_3/R_2)}{1 + sCR_1} \qquad (7.31)$$

Single-ended input realizations (Fig. 7.10(b)) have also been suggested[15] as well as a dual-output circuit.[16] These are all active versions of the familiar phase-splitting *LC* lattice networks. The available phase shift may be increased to 360° without employing a transformer by connecting a Wien-bridge[17] or twin-T circuit replacing $(Y_A - Y_C)$ at the non-inverting input of the amplifier (Fig. 7.10(c)).

Such allpass sections have found extensive application as basic blocks in low-sensitivity biquadratic filters,[18,19] circulators,[20] and waveform correctors.

7.3.3 Modified multiple-loop feedback circuits

In order to relieve the necessity for wide component spreads in the attainment of high Q's with the basic multiple-loop feedback filter, positive feedback may be introduced via the non-inverting input of the DIOA. This allows independent control of Q and centre frequency. Further generalization leads to the SAB (single-amplifier biquadratic) section of the form shown in Fig. 7.11

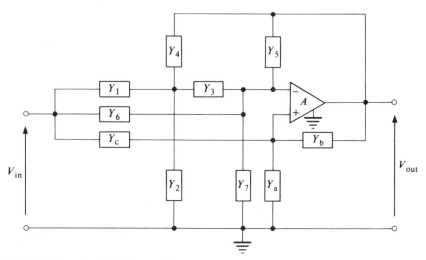

Fig. 7.11. Generalized SAB MLFB filter

with admittances numbered in the negative-feedback path and lettered in the positive-feedback path. Analysis yields the transfer ratio

$$\frac{V_{out}}{V_{in}} = \frac{\begin{Bmatrix} Y_c Y_3 (Y_1 + Y_2 + Y_4) + Y_c (Y_5 + Y_6 + Y_7) \\ \times (Y_1 + Y_2 + Y_3 + Y_4) - (Y_a + Y_b + Y_c) \\ \times (Y_1 Y_3 + Y_6 (Y_3 + Y_4) + Y_6 (Y_1 + Y_2)) \end{Bmatrix}}{\begin{Bmatrix} (Y_a + Y_c)(Y_3 Y_4 + Y_5(Y_1 + Y_2 + Y_3 + Y_4)) \\ - Y_b (Y_3(Y_1 + Y_2) + (Y_6 + Y_7) \\ \times [Y_1 + Y_2 + Y_3 + Y_4]) + (1/A)(Y_a + Y_b + Y_c) \\ \times ([Y_5 + Y_6 + Y_7](Y_1 + Y_2 + Y_3 + Y_4) \\ + Y_3(Y_1 + Y_2 + Y_4)) \end{Bmatrix}} \quad (7.32)$$

The negative terms in the numerator associated with the additional positive-path admittances Y_a, Y_b, and Y_c obviously allows realization of positive finite-zero transfer functions. The Friend SAB circuit[21] emerges with the substitutions $Y_1 = G_1'$, $Y_2 = G_2$, $Y_3 = sC_3$, $Y_4 = sC_4$, $Y_5 = G_5$, $Y_6 = G_6$, $Y_7 = G_7$, $Y_a = G_a$, $Y_b = G_b$, $Y_c = G_c$ and, for an ideal infinite-gain operational amplifier, design equations have been developed using C_3, C_4, $(G_a + G_c)$, and G_b as the dependent variables:

$$G_1 + G_2 = \frac{C_4(G_a + G_c)}{2G_b} \left\{ -b_1 + b_1^2 + 4\left(1 + \frac{C_3}{C_4}\right) b_0 \left(\frac{G_b}{G_a + G_c}\right) \right\}^{1/2}$$

(7.33a)

$$G_1 = \frac{[(G_1 + G_2)a_2 + (1 + C_3/C_4)a_0 [C_4^2/(G_1 + G_2)] - a_1 C_4]}{1 + G_b/(G_a + G_c)}$$

(7.33b)

$$G_6 + G_7 = \frac{C_3 C_4 (G_a + G_c)(a_0 - b_0 a_2)}{(G_1 + G_2)(G_a + G_b + G_c)[a_2 - G_6/(G_6 + G_7)]}$$

(7.33c)

$$G_5 = \frac{C_3 C_4 b_0}{G_1 + G_2} + \frac{G_b (G_6 + G_7)}{G_a + G_c}$$

(7.33d)

$$G_c = a_2 (G_a + G_c)$$

(7.33e)

Various special cases can be obtained, e.g., the highpass notch (with $G_7 = 0$), lowpass notch (with $G_6 = 0$), and lowpass (with $Y_2 = sC_2$, $Y_3 = G_3$, $Y_4 = G_4$, $Y_6 = Y_7 = Y_c = 0$). When large resistance values seem necessary, introduction of phantom resistors and T–π conversions produce the STAR configurations for integrated-circuit implementation.

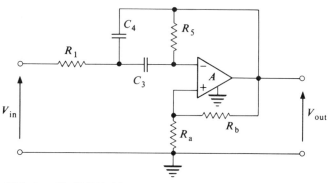

Fig. 7.12. High-Q modified MLFB filter

A particularly convenient form arises by the choice $Y_2 = Y_6 = Y_7 = Y_c = 0$, giving

$$\frac{V_{out}}{V_{in}} = \frac{-Y_1 Y_3 (1+k)}{Y_5(Y_1 + Y_3 + Y_4) + Y_3(Y_4 - kY_1)} \quad (7.34)$$

where $k = G_b/G_a$. This includes the Deliyannis[22] high-Q bandpass filter of the form of Fig. 7.12 and with transfer function

$$\frac{V_{out}}{V_{in}} = \frac{-sC_3 G_1 (1+k)}{s^2 C_3 C_4 + s\{G_5(C_3 + C_4) - kC_3 G_1\} + G_1 G_5} \quad (7.35)$$

for which design proceeds by coefficient matching.

7.4 Optimum design and sensitivity

The synthesis methods described in earlier sections are summarized in Fig. 7.13, which serves as an initial design guide. The material already presented furnishes a certain amount of performance data, particularly in relation to design facility and component count. For instance, while the DIOA general configurations require as many as eleven components for all functions except the Wien allpass of Fig. 7.10(c), the other structures offer significant savings for all-pole realizations. The single-ended-input multiple-loop feedback (MLFB) network is most economical in this respect, requiring only five passive elements, though complex zeros necessitate additional elements as part of a more sophisticated synthesis technique.

More comprehensive comparisons can only be made on the basis of a common transfer function, the bandpass being adopted in a manner similar to that of the previous chapter (Sec. 6.4). The Q and ω_0 sensitivities presented in Table 7.3 are evaluated by means of Eqs. (4.33) and (4.34), the technique used and assumptions made now being demonstrated for the case of multiple-loop

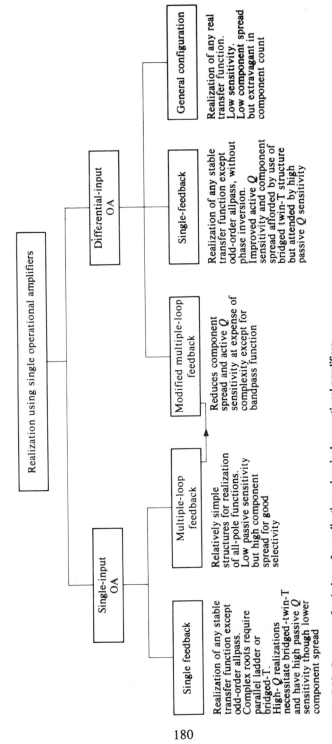

Fig. 7.13. Summary of techniques for realization using single operational amplifiers

REALIZATION USING SINGLE OPERATIONAL AMPLIFIERS

Performance parameters for bandpass transfer function

Network structure		Q	ω_0	S_Q^ω	$S_A^{\omega_0}$	$\|S_{RC}^Q\|_{max}$	$\|S_{RC}^{\omega_0}\|_{max}$	Maximum component spread	Design conditions
Single-ended-input single feedback (Fig. 7.1)	Bridged T	$\dfrac{\sqrt{C_2/C_1}}{\sqrt{G_1/G_2}+\sqrt{G_2/G_1}}$	$\sqrt{\dfrac{G_1 G_2}{C_1 C_2}}$	$\dfrac{Q(2Q+H)}{A}$	0	$\dfrac{1}{2}$	$\dfrac{1}{2}$	$4Q^2$	$C_1 = C_2/4Q^2 = C/HQ$; $G_1 = G_2 = G/H$
	Bridged twin-T	$\dfrac{G_1}{2G_0}$	$\dfrac{G_1}{C_1}$	$\dfrac{6Q}{A}$	0	$\dfrac{Q}{2}$	$\dfrac{3}{8}$	$4Q$	$C_1 = C_2 = C_3/2 = C/H$; $G_1 = G_2 = G_3/2 = G/H = 2QG_0$; $Q \gg 1$
Multiple-loop feedback (Fig. 7.5)		$\dfrac{\sqrt{(G_1+G_2)/G_3}}{\sqrt{C_3/C_4}+\sqrt{C_4/C_3}}$	$\sqrt{\dfrac{G_3(G_1+G_2)}{C_3 C_4}}$	$\dfrac{2Q^2}{A}$	0	$\dfrac{1}{2}$	$\dfrac{1}{2}$	$2Q^2$	$C_3 = C_4$, $G_1 = G_2 = 2Q^2 G_3$; $H = Q\omega_0$
Differential-input single feedback (Fig. 7.6)	Bridged T	$\dfrac{\sqrt{C_2/C_1}}{\sqrt{G_1/G_2}+\sqrt{G_2/G_1}}$	$\sqrt{\dfrac{G_1 G_2}{C_1 C_2}}$	$\dfrac{2Q^2}{A}$	0	$\dfrac{1}{2}$	$\dfrac{1}{2}$	$4Q^2$	$C_1 = C_2/4Q^2 = C_4/4Q^2 = C_3/4Q^2$; $G_1 = G_2 = \dfrac{G_4}{\sqrt{2QH}} = \dfrac{\sqrt{2QH}}{4Q^2} \cdot G_3$
	Bridged twin-T	$\dfrac{G_1}{2G_0}$	$\dfrac{G_1}{C_1}$	$\dfrac{4Q}{A}$	0	$\dfrac{Q}{2}$	$\dfrac{3}{8}$	$4Q$	$C_1 = C_2 = C_3/2 = C_4$ $= C_3/(4H-H^2-1)$; $G_1 = G_2 = G_3/2 = G_4$ $= G_5/(4-H-1/H) = 2QG_0$; $Q \gg 1$
Differential-input general configuration (Fig. 7.8)		$\dfrac{\sqrt{G_F/G_E}}{\sqrt{C_F/C_E}+(G_F/G_E-1)\sqrt{C_E/C_F}}$	$\sqrt{\dfrac{G_E G_F}{C_E C_F}}$	$\dfrac{5Q}{A}$	0	$\dfrac{Q}{2}$	$\dfrac{1}{2}$	$1+\dfrac{2}{H}$	$C_B = G_B = 0$; $C_A = HC_C = HC_F = H\dfrac{Q}{2Q-1} \cdot C_E$ $= H\dfrac{Q}{(2+H)Q-1} \cdot C_D$; $G_A = HG_C = \ldots$, as for C's; $Q \gg 1$
Modified multiple-loop feedback (Fig. 7.12)		$\dfrac{\sqrt{G_1/G_3}}{\sqrt{C_3/C_4}+\sqrt{C_4/C_3}-k(G_1/G_3)\sqrt{C_3/C_4}}$	$\sqrt{\dfrac{G_1 G_3}{C_3 C_4}}$	$\dfrac{Q}{A\sqrt{r}}$	0	$2Q\sqrt{r}-\dfrac{1}{2}$	$\dfrac{1}{2}$	r	$r = G_3/G_1$, $C_3 = C_4$; $k = 2r - \sqrt{r}/Q$, $H = (1+k)/\sqrt{r}$ $= G_b/G_a$

analysis. Taking account of the finite amplifier gain A but assuming infinite input impedance and zero output impedance (i.e., $Y_{11} = Y_{44} = 0$ in Fig. 5.5(c)), the bandpass structure of Table 7.1 analyses to the voltage transfer function

$$\frac{V_2}{V_1} = \frac{-\dfrac{G_1}{\left(1 + \dfrac{1}{A}\right)C_4}s}{s^2 + \left[\dfrac{G_5(C_3 + C_4)}{C_3 C_4} + \dfrac{G_1 + G_2}{(1 + A)C_4}\right]s + \left[\dfrac{G_5(G_1 + G_2)}{C_3 C_4}\right]} \quad (7.36)$$

which is identical to that in Table 7.1 when A is infinite. The actual Q factor is given by

$$Q_k = \frac{Q}{1 + C_3(G_1 + G_2)/G_5(C_3 + C_4)(1 + A)} \quad (7.37)$$

where $Q = Q_k|_{A \to \infty}$ is the design value listed in Table 7.3. For a maximum-Q design, the element values of Sec. 7.2 are pertinent, reducing Eq. (7.37) to

$$Q_k = \frac{(1 + A)Q}{(1 + A) + 2Q^2} \quad (7.38)$$

Applying the classical definition of Eq. (4.33) produces

$$S_A^{Q_k} = \frac{2Q^2 A}{(1 + A)[1 + A + 2Q]} \to \frac{2Q^2}{A} \quad (7.39)$$

if $A \gg 2Q + 1$, as listed in Table 7.3. It is common practice to omit the suffix k when discussing Q sensitivity. The sensitivities of Q to passive elements are given by Eq. (4.33) as

$$S_{C_3, C_4}^Q = 0, \quad S_{G_5}^Q = -0.5, \quad \text{and} \quad S_{G_1, G_2}^Q = \frac{G_1}{2(G_1 + G_2)} = 0.25$$

since $G_1 = G_2$, the maximum modulus of 0.5 being listed in Table 7.3. Similar reasoning holds for the frequency sensitivity to passive elements, that to A clearly being zero by the form of Eq. (7.36). The column for the maximum component spread presents the conductance or capacitance ratio, whichever is the greater. In the case of the multiple-loop network, the Q dependence dictates that it is the conductance ratio.

Similar analysis and application of the techniques of Chapter 4 where appropriate leads to the essential results presented in Table 7.3. It should be remembered that the simplified expressions listed for Q and ω_0 are design

REALIZATION USING SINGLE OPERATIONAL AMPLIFIERS

values only and the corresponding sensitivity results do not necessarily follow directly from them. The importance of deriving sensitivity from the general expression is exemplified by considering the single-feedback bridged twin-T entry for ω_0, listed as G/C and for which one would expect to evaluate $S_G^{\omega_0} = -S_C^{\omega_0} = 1$. However, the true value for ω_0 sensitivity must be determined before the balanced twin-T element relationships are included and necessitates operation on a third-order expression. The maximum value obtained is

$$S_{C_1}^{\omega_0} = \frac{-(3Q-1)}{4(2Q-1)} \rightarrow -0.375$$

(with an error <2 per cent when $Q = 10$) which is the value presented in the table as an initial design guide. Also, the expression for Q sensitivity to amplifier gain is shown as $6Q/A$ rather than $(6Q + 1)/A$ under the assumption that $6Q \gg 1$ stipulated under the design conditions.

All bandpass realizations listed in Table 7.3 can have at least one pair of equal resistors and one pair of equal capacitors. Indeed the multiple-loop configuration has only one unequal normalized element value. However, this apparent advantage is lost when account is taken of the resistor spread of Q^2 proportions. This is a considerable disadvantage for microelectronic fabrication, possibly necessitating an additional discrete component. As the table indicates, there is a trade-off between maximum component spread and passive Q sensitivity. The bridged-T and its multiple-loop feedback equivalent have low sensitivity but high component spread (witness $R_1 = 1/2Q$ in Ex. 7.1), while the twin-T configurations have a more tolerable range of component values at the expense of their notorious passive sensitivity. It is interesting to observe the effect of using the differential-input facility of the operational amplifier: for both twin-T and bridged-T configurations, Q sensitivity is reduced and in the case of the latter the dependence on gain constant H removed. On a component-spread basis alone, the general differential-input configuration appears to be optimum, but this is offset by its extravagant component count.

The most flexible design is that afforded by the modified multiple-loop network whose tabulated results merit further comment. Optimum design ensues by selecting[22] equal-valued capacitors. Setting $C_3 = C_4$, coefficient matching together with Eq. (7.35) yields the design equation

$$k = 2r - \sqrt{r}/Q$$

where $r = G_5/G_1$. Since k must be a positive-real number, $r > 1/4Q^2$. Although not affecting ω_0 sensitivity, r constitutes a control parameter for both Q sensitivity and component spread as indicated in Table 7.3. Choosing r close to its limit gives low values for one at the expense of the other.

References

1. *Handbook of Operational Amplifier Active RC Networks*, Burr-Brown Research Corporation, Arizona, 1966, Chap. 2, pp. 15–26.
2. N. Balabanian, *Network Synthesis*, Prentice Hall, Englewood Cliffs, N.J., 1958, Chap. 7.
3. S. K. Mitra, *Analysis and Synthesis of Linear Active Networks*, Wiley, New York, 1969, Appendix D, pp. 547–553.
4. L. P. Huelsman, *Theory and Design of Active RC Circuits*, McGraw-Hill, New York, 1968, Chap. 6, p. 189.
5. G. S. Moschytz, 'A general approach to twin-T design and its application to hybrid integrated linear networks', *Bell System Tech. J.*, **49**, 1105–1149, 1970.
6. E. V. Bohn, *Transform Analysis of Linear Systems*, Addison-Wesley, Reading, Mass., 1963, p. 71.
7. G. K. Aggarwal, 'On nth order simulation by one operational amplifier', *Proc. IEEE*, **52**, 969, 1969.
8. P. L. Taylor, 'Flexible design method for active RC two-ports', *Proc. IEE*, **110**, 1607–1616, 1963.
9. A. G. J. Holt and J. I. Sewell, 'Active RC filters employing a single operational amplifier to obtain biquadratic responses', *Proc. IEE*, **112**, 2227–2234, 1965.
10. M. H. Nichols and L. Rauch, *Radio Telemetry*, Wiley, New York, 1956, p. 396.
11. A. G. J. Holt and J. I. Sewell, 'Synthesis of multiple loop feedback systems—Part II', *Int. J. Control*, **9** (5), 499–508, 1969.
12. T. A. Hamilton and A. S. Sedra, 'Some new configurations for active filters', *IEEE Trans.*, **CT-19** (1), 25–33, 1972.
13. J. E. B. Ponsonby, 'Active all-pass filter using a differential operational amplifier', *Electron. Letters*, **2** (4), 134–135, 1966.
14. I. F. Macdiarmid, 'Active all-pass filter using a differential amplifier', *Electron. Letters*, **2** (5), 186, 1966.
15. P. Aronhime and A. Budak, 'An operational all-pass network', *Proc. IEEE*, **57**, 1677–1678, 1969.
16. R. Tarmy and M. S. Ghausi, 'Very high-Q insensitive active RC networks', *IEEE Trans.*, **CT-17** (3), 358–366, 1969.
17. R. Genin, 'Realization of an all-pass transfer function using operational amplifiers', *Proc. IEEE*, **56**, 1746–1747, 1968.
18. D. J. Comer and J. E. McDiarmid, 'Inductorless bandpass characteristics using all-pass networks', *IEEE Trans.*, **CT-15**, 501–502, 1968.
19. G. Wilson, Y. Bedri, and P. Bowron, 'RC-active networks with reduced sensitivity to amplifier gain-bandwidth product', *IEEE Trans.*, **CAS-21** (5), 618–626, 1974.
20. A. W. Keen, J. L. Glover, and R. J. Harris, 'Realization of the circulator concept using differential-input operational amplifier', *Electron. Letters*, **4** (18), 389–391, 1968.
21. J. J. Friend, C. A. Harris, and D. Hilberman, 'STAR: An active biquadratic filter section', *IEEE Trans.*, **CAS-22** (2), 115–121, 1975.
22. T. Deliyannis, 'High-Q factor circuit with reduced sensitivity', *Electron. Letters*, **4** (26), 577–579, 1968.

8. Realization by analogue simulation

The various active elements used in the realization of inductorless filters have been presented in Chapter 5. Network structures incorporating controlled sources were the subject of Chapter 6 while Chapter 7 considered the direct application of the operational amplifier as the active element. In this chapter, the realization methods presented are based mainly on the simulation of the well-established passive-LC ladder structure[1] by means of converters, inverters, and circulators, all of which were introduced in Chapter 5. The attractions of this approach will become apparent below.

8.1 Direct replacement of inductors

The implementation of active filters by direct replacement of inductors has some distinct advantages over other techniques. In the first instance, the designer is able to avail himself of existing passive-design tables,[2] thereby being relieved of the task of determining element values from fundamentals. Second, the resulting active filter retains the low-sensitivity properties of the passive ladder from which it was derived. This latter point cannot be overstressed since so many other active filters exhibit sensitivities considerably greater than their passive counterparts (see Chapters 6 and 7). However, the advantages of the technique are mainly limited to highpass and bandpass ladder structures with their grounded inductors and inherent constraints. Direct substitution of floating coils in lowpass and bandpass ladders as well as other LC configurations is not so straightforward (see Sec. 8.1.2).

8.1.1 Grounded inductor simulation

Practical inductance coils are not ideal elements and have finite quality factors, as pointed out in Chapter 4. These values are usually less than 1000 in the audio band and much lower at higher frequencies. Hence, although simulated inductors can be represented by the models shown in Fig. 4.2, the active replacement will usually be at least as good as the original coil and the simulation of actual lossy elements is not necessary for most applications.

The realization of grounded inductors is relatively straightforward and well-established. Though they may be directly simulated using operational amplifiers or a GIC, the most widely accepted technique is the capacitive termination of a gyrator. It is clear from Chapter 5 that active realizations of ideal building blocks (NIC, NII, etc.) contain unwanted terms. The gyrator is no exception to this, as seen from Eq. (5.35). Table 8.1 illustrates two dual representations of a non-ideal gyrator together with their matrix descriptions. If port 2 is terminated by an ideal capacitor, C, the resulting simulated inductance can be calculated as a series or parallel model. The results provide a means of assessing the quality of inductance simulated by various gyrator realizations.

Table 8.1 Non-ideal representations of gyrator

No.	Model Circuit	Matrix description
1	(circuit with G_1, $\pm G_a V_2$, $\mp G_b V_1$, G_2)	$\begin{bmatrix} I_1 \\ I_2 \end{bmatrix} = \begin{bmatrix} G_1 & \pm G_a \\ \mp G_b & +G_2 \end{bmatrix} \begin{bmatrix} V_1 \\ V_2 \end{bmatrix}$
2	(circuit with R_1, $V_1 \pm R_a I_2$, $\mp R_b I_1$, R_2)	$\begin{bmatrix} V_1 \\ V_2 \end{bmatrix} = \begin{bmatrix} R_1 & \pm R_a \\ \mp R_b & R_2 \end{bmatrix} \begin{bmatrix} I_1 \\ I_2 \end{bmatrix}$

As an example, consider the series inductor produced by capacitively terminating Model 1 of Table 8.1. The composite transmission matrix may be derived by conversions (Appendix 5) as

$$[a] = \begin{bmatrix} \dfrac{G_2 + sC}{G_b} & \dfrac{1}{G_b} \\ \dfrac{G_1 G_2 + G_a G_b + sCG_1}{G_b} & \dfrac{G_1}{G_b} \end{bmatrix} \tag{8.1}$$

from which the input impedance is

$$Z_{in} = \dfrac{a_{11}}{a_{21}} = \dfrac{G_2 + sC}{G_1 G_2 + G_a G_b + sCG_1} \tag{8.2}$$

Table 8.2 Expressions for inductance simulated by capacitively terminated gyrator

Element or parameter		Model 1		Model 2	
		Series	Parallel	Series	Parallel
L	Exact	$\dfrac{CG_aG_b}{(G_1G_2+G_aG_b)^2+\omega^2C^2G_1^2}$	$\dfrac{G_2^2+\omega^2C^2}{\omega^2CG_aG_b}$	$\dfrac{CR_aR_b}{1+\omega^2C^2R_2^2}$	$\dfrac{R_1^2+\omega^2C^2(R_1R_2+R_aR_b)^2}{\omega^2CR_aR_b}$
	Ideal		C/G^2		CR^2
R	Exact	$\dfrac{G_2(G_1G_2+G_aG_b)+\omega^2C^2G_1}{(G_1G_2+G_aG_b)^2+\omega^2C^2G_1^2}$	$\dfrac{G_2^2+\omega^2C^2}{(G_1G_2+G_aG_b)G_2+\omega^2C^2G_1}$	$\dfrac{R_1+\omega^2C^2R_2(R_1R_2+R_aR_b)}{1+\omega^2C^2R_2^2}$	$\dfrac{R_1^2+\omega^2C^2R_2(R_1R_2+R_aR_b)}{R_1+\omega^2C^2R_2(R_1R_2+R_aR_b)}$
Q_s	Exact		$\dfrac{\omega CG_aG_b}{G_2(G_1G_2+G_aG_b)+\omega^2C^2G_1}$		$\dfrac{\omega CR_aR_b}{R_1+\omega^2C^2R_2(R_1R_2+R_aR_b)}$
$\hat{\omega}$	Exact		$\dfrac{G_2}{C_2}\left(1+\dfrac{G_aG_b}{G_1G_2}\right)^{1/2}$		$\left\{C^2R_2^2\left(1+\dfrac{R_aR_b}{R_1R_2}\right)\right\}^{-1/2}$
	Ideal		G/C		$1/CR$
$\hat{Q}_s$	Exact		$\dfrac{G_aG_b}{2G_1G_2}\left(1+\dfrac{G_aG_b}{G_1G_2}\right)^{-1/2}$		$\dfrac{R_aR_b}{2R_1R_2(1+R_aR_b/R_1R_2)^{1/2}}$
	Ideal		$G/2G_1$		$R/2R_1$

187

Note that for the perfect symmetrical gyrator ($G_1 = G_2 = 0$, $G_a = G_b = G$), $Z_{in} = sC/G^2$, in accordance with Eq. (5.33). Equation (8.2) may be rationalized as

$$Z_{in}(j\omega) = R_s + j\omega L_s = \left[\frac{G_2(G_1G_2 + G_aG_b) + \omega^2 C^2 G_1}{(G_1G_2 + G_aG_b)^2 + \omega^2 C^2 G_1^2}\right]$$
$$+ j\omega \left[\frac{CG_aG_b}{(G_1G_2 + G_aG_b)^2 + \omega^2 C^2 G_1^2}\right]$$

The complete expressions for R_s, L_s and Q_s are presented in Table 8.2, together with those for optimum quality factor ($\hat{Q}_s$) and the radian frequency at which it occurs ($\hat{\omega}$). Under the assumptions indicated in the table, $\hat{Q}_s$ can be evaluated[3,4] as $G/2G_1$, occurring at $\hat{\omega} = G/C$. The remaining entries in Table 8.2 may be obtained in a fashion similar to that outlined above. The table can be used to evaluate the performance of a gyrator if either its admittance or impedance parameters are known.

Example 8.1 Use the Antoniou gyrator described in Fig. 5.22(c) to simulate a 1 H inductor across port 1 by terminating port 2 with a 1 µF capacitor. Regard the amplifiers as a matched pair having an open-loop gain of 10^4 and assume equal-valued resistors throughout. Calculate the resistor values for the gyrator, the optimum quality factor and the frequency at which it occurs.

SOLUTION The impedance parameters for the gyrator as presented in Eq. (5.35) can be equated to those in Table 8.1 to give

$$Z_{11} = \frac{R(R'_1 + R)}{R'_1 + R(1 + A)} = R_1 \quad (8.3a)$$

$$Z_{12} = \frac{R'_2(R'_1 + RA)}{R'_1 + R(1 + A)} = R_a \quad (8.3b)$$

$$Z_{21} = \frac{R(R + R'_1A)}{R'_1 + R(1 + A)} = R_b \quad (8.3c)$$

$$Z_{22} = \frac{R'_2(R'_1 + R)}{R'_1 + R(1 + A)} = R_2 \quad (8.3d)$$

where $A_1 = A_2 = A$ and the original values of R_1 and R_2 have been changed to R'_1, R'_2 so as to avoid conflict of terminology.

Setting $R'_1 = R'_2 = R$ is a desirable objective and has no obvious practical disadvantage. (In the nullator–norator model, it sets $a_{12} = -a_{21}$ in the NII.) As

REALIZATION BY ANALOGUE SIMULATION

a direct consequence, $R_a = R_b = R$ and $R_1 = R_2 = 2R/A$, thereby representing the ideal case listed in Table 8.2. Thus, the simulated inductance is given by

$$L = CR^2$$

and, for $L = 1$ H, $C = 1$ μF implies $R = 1$ kΩ. Hence, $\hat{\omega} = 1/CR = 10^3$ rad s^{-1}. The optimum quality factor may be obtained from Table 8.2 as

$$\hat{Q}_s = \frac{R}{2R'} = \frac{A}{4} = 2500$$

The above example is realistic; quality factors in excess of 2000 at 200 Hz have been achieved when simulating[5] a 1 H inductor, while two-amplifier gyrators[6] can give values of $\hat{Q}_s$ as high as $A/2$.

8.1.2 Simulation of floating inductors

The presence of floating or non-grounded inductors in lowpass and bandpass LC filters makes their simulation necessary in direct-replacement methods. The gyrators discussed so far have all been three-terminal devices and their use (as single elements) in floating-inductor simulation would necessitate the use of floating power supplies or a gyrator-flotation circuit.[7] An alternative is the utilization[8] of fully-floating gyrators symbolized in Fig. 8.1(a) though, in practice, it is usually possible to ground the capacitively loaded port as in Fig. 8.1(b), thereby creating[9] a semi-floating gyrator. Most realizations of floating-port gyrators are known to be equivalent to the arrangement of Fig. 8.2, which comprises[10] a cascade of two grounded gyrators and an embedded capacitor.

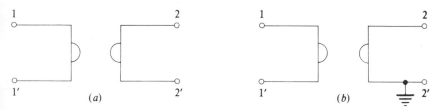

Fig. 8.1. Gyrator symbols. (a) Fully-floating; (b) semi-floating

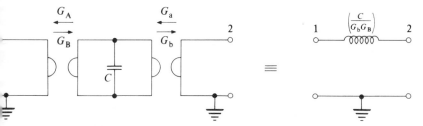

Fig. 8.2. Realization of floating inductance using grounded gyrators

The composite transmission matrix may be obtained from Table 5.6 as follows:

$$[a]_{comp} = [a]_{gyrator\ 1} \times [a]_{capacitor} \times [a]_{gyrator\ 2}$$

$$= \begin{bmatrix} 0 & \dfrac{1}{G_B} \\ G_A & 0 \end{bmatrix} \times \begin{bmatrix} 1 & 0 \\ sC & 1 \end{bmatrix} \times \begin{bmatrix} 0 & \dfrac{1}{G_b} \\ G_a & 0 \end{bmatrix}$$

$$= \begin{bmatrix} \dfrac{G_a}{G_B} & \dfrac{sC}{G_b G_B} \\ 0 & \dfrac{G_A}{G_b} \end{bmatrix} \quad (8.4)$$

the gyrators having unequal conductances but being otherwise ideal. Should $G_a = G_B$ and $G_A' = G_b$, then it represents the transmission matrix of a floating inductor of value $(C/G_b G_B)$. For many applications, this technique is perfectly adequate. However, for complex filters, the effect of gyration-conductance mismatch may be significant. Indeed, Eq. (8.4) actually represents a more complex inductive-π arrangement and it has been shown[7] that as little as a two per cent variation in one of the gyrator conductances can cause noticeable degradation in the passband response of a 16th-order bandpass section.

An alternative realization method employing two GIC's and a floating resistor is shown in Fig. 8.3. Using identical converters, the composite transmission matrix is

$$[a] = \begin{bmatrix} 1 & 0 \\ 0 & 1/sk \end{bmatrix} \begin{bmatrix} 1 & R \\ 0 & 1 \end{bmatrix} \begin{bmatrix} 1 & 0 \\ 0 & sk \end{bmatrix} = \begin{bmatrix} 1 & skR \\ 0 & 1 \end{bmatrix}$$

which is clearly that of a floating inductance of value kR.

Multiport circulators may also be used to simulate floating inductance[11] and have the advantage of using fewer amplifiers than the gyrator cascade method, assuming a two-amplifier realization of each gyrator. Figure 8.4 shows a lossless three-port circulator (following Sec. 5.4.5) in which all transfer admittances are real and equal to G. Using the admittance matrix presented in

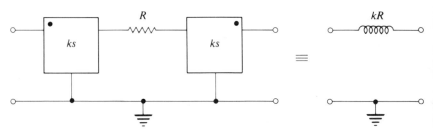

Fig. 8.3. Realization of floating inductance using grounded GIC's

REALIZATION BY ANALOGUE SIMULATION

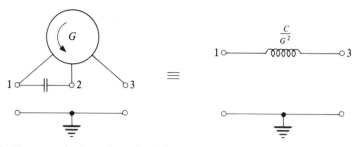

Fig. 8.4. Three-port circulator simulation of floating inductance

Eq. (5.40), the composite matrix, due to the presence of the capacitive load, becomes

$$[Y]_{comp} = \begin{bmatrix} sC & (G-sC) & -G \\ -(G+sC) & sC & G \\ G & -G & 0 \end{bmatrix}$$

Pivotal condensation about node 2 leads to the matrix

$$[Y] = \frac{G^2}{sC} \begin{bmatrix} 1 & -1 \\ -1 & 1 \end{bmatrix}$$

which can be represented by a non-grounded inductor of value $L = C/G^2$. Capacitively terminated, lossless three-port circulators may thus be used as direct replacements for non-grounded inductors. Indirect replacement is also possible using higher-order circulators as is discussed briefly in the following subsection.

8.1.3 Active-ladder structures

Many texts are available on the theory of *RLC* synthesis and it is the purpose of this subsection merely to present a selection of commonly used filter sections and comment on their suitability as active-ladder prototypes. All passive filters can be designed from normalized lowpass sections, examples of which are shown in Table 8.3. Transformation to highpass structures is effected simply by replacing inductors (capacitors) by capacitors (inductors) having reciprocal values. The bandpass and bandstop sections are derived in a less simple fashion but may, nevertheless, be effected[2] without undue labour. All sections in Table 8.3 can be realized electronically by directly replacing the inductors by their simulated equivalents. It is apparent that lowpass sections always contain floating inductors while the other types usually have at least some grounded coils. It is often possible[4] either to reduce the number of floating inductors to be replaced or, in some cases, to eliminate them altogether.

Table 8.3 Examples of *RLC* sections for ladder structures

Lowpass prototype	Transformed section	
	Bandpass	Bandstop
(a) All-pole		
(b) Elliptic		

Another method[11] of simulating floating inductance may be regarded as a form of semi-direct realization in which the number of circulators is less than the number of replaced coils. It makes use of a four-port circulator in which there is phase reversal between two of the ports. It can be shown that the two circuits in Fig. 8.5 (in which port phase reversal is indicated by an asterisk) are equivalent, so that the 4-port circulators and attendant capacitors have relaced three floating inductors. Although there is not a direct topological equivalence between the two networks, the non-converted element values are equal. Thus, C_1, C_3, C_5, C_7, R_S, and R_L have the same values in each case while $C_2 = L_2 G^2$,

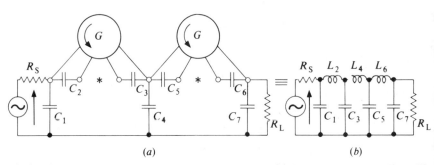

Fig. 8.5. Inductor replacement using four-port circulators. (*a*) Seventh-order active filter; (*b*) seventh-order passive filter

REALIZATION BY ANALOGUE SIMULATION

$C_4 = L_4 G^2$, and $C_6 = L_6 G^2$. This approach can be extended to higher-order filters but is only effective in terms of amplifier savings for low-order sections.

8.2 Analogue methods

In contrast to the above method of 'direct' replacement, analogue methods are essentially 'indirect' realizations. It may be argued that such an approach is more suited to inclusion in Chapter 7. However, the preceding chapter is devoted to the realization of second-order filter stages using single and differential-input operational amplifiers, whereas the present theme is the realization of high-order stages by analogy with passive-component designs. Hence, analogous design, not being limited by second-order-cascade restraints, has probably a stronger association with this chapter. This link will become more evident below where the two philosophies of leap-frogging and state variables are discussed.

8.2.1 Leapfrog filter

This method of active synthesis relies on simulation of the interactive effects in a ladder structure by feedback loops as in a control system. The passive reactances may then be replaced on a one-to-one basis by Miller[12] integrators. As such, it is particularly suited to the realization of lowpass responses which fortuitously happen to be the most difficult to produce by the methods of Sec. 8.1. The importance of this early[13] procedure is only now being appreciated in the context of the manifest advantages of multi-loop *vis à vis* cascade techniques (Sec. 9.6).

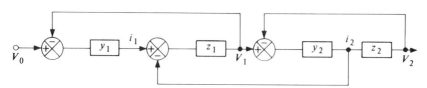

Fig. 8.6. Multiple-loop feedback system

Consider the analogy between the variables in the first two sections of the immittance ladder of Fig. 3.11 and those in the multiloop feedback system of Fig. 8.6. Each may be described by the set of equations

$$V_2 = i_2 Z_2 \qquad (8.5a)$$
$$i_2 = (V_1 - V_2) y_2 \qquad (8.5b)$$
$$V_1 = (i_1 - i_2) Z_1 \qquad (8.5c)$$
$$i_1 = (V_0 - V_1) y_1 \qquad (8.5d)$$

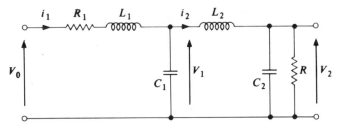

Fig. 8.7. Lowpass ladder network

Thus the feedback system may be constructed using analogue computer principles, the analogue variables clearly being all voltages, representing both voltages and currents in the original ladder. Each transfer-function block in Fig. 8.6 represents an immittance in Fig. 3.11, where $Z_i = 1/Y_i$ and $y_i = 1/z_i$.

Simulation of the lowpass ladder network shown in Fig. 8.7 will serve to illustrate the technique.[13] The voltages and currents are consistent with those in Fig. 8.6 and, in terms of immittances,

$$y_1 = \frac{1}{(R + sL_1)}, \quad Z_1 = \frac{1}{sC_1}, \quad y_2 = \frac{1}{sL_2}, \quad \text{and} \quad Z_2 = \frac{1}{(sC_2 + 1/R)}$$

If the ladder is normalized with respect to the terminating resistance, R, the immittances become dimensionless and may be represented by the gain blocks in Fig. 8.6. Thus, all admittances are multiplied by R and all impedances are divided by R. That this does not affect the voltage transfer function is seen by combining Eqs. (8.5a–d) into

$$\frac{V_2}{V_0} = \frac{Z_1 Z_2 y_1 y_2}{1 + Z_1 y_1 + Z_1 y_2 + Z_2 y_2 + Z_1 Z_2 y_1 y_2}$$

$$= \frac{1}{1 + z_1 Y_1 + z_2 Y_2 + z_1 Y_2 + z_1 z_2 Y_1 Y_2} \quad (8.6)$$

which is consistent with the cumulant of Eq. (3.32) and is clearly unaffected by the normalization process.

The analogue realizations of the four basic gain blocks are presented in Table 8.4, together with details of the element values. The basic block, as for an analogue computer, is the integrator, and there is a simple time-constant relationship between the components in the passive ladder and those in the analogue system. In addition to the gain blocks, each feedback loop requires an inverting amplifier of the form described in Chapter 5. If these inverters are placed in the forward path between input and output, fewer will be required than if they are placed in the outer feedback loops. Thus, in Fig. 8.6, two

REALIZATION BY ANALOGUE SIMULATION

Table 8.4 Realization of analogue function blocks for lowpass ladder simulation

Ladder element	Normalized immittance	Active realization	Active transfer function	Analogue symbol
y_1	$\dfrac{(R/R_1)}{1 + sT_1'(R/R_1)}$		$\dfrac{-(R_f/R)}{1 + sT_1'(R_f/R)}$	y_{1L}
Z_1	$\dfrac{1}{sT_1}$		$\dfrac{-1}{sT_1}$	T_1
y_2	$\dfrac{1}{sT_2'}$		$\dfrac{-1}{sT_2'}$	T_2'
Z_2	$\dfrac{1}{sT_2 + 1}$		$\dfrac{-1}{sT_2 + 1}$	Z_{2L}

$T_i = C_i R,\ T_i' = L_i/R = C_i' R,\ R_f = R^2/R_1.$

inverters will suffice if placed immediately after Z_1 and Z_2 (cf., three if placed in the feedback paths of each loop). Alternatively, the integrator-inverter cascade may be replaced by a non-inverting integrator, preferably[14] of the two-amplifier variety. High-order filters can be designed by pattern repetition and lowpass ladders having a shunt capacitor as the first reactive element are easily accommodated.

Example 8.2 Use the leapfrog method to design a fifth-order lowpass Gaussian filter (Chapter 2) having equal terminating resistors and 3 dB loss at $\omega = 1$ rad s^{-1}.

SOLUTION The passive ladder is shown in Fig. 8.8(a), the element values having been obtained from standard tables.[15] Immittances of the required form may be identified as indicated in Fig. 8.8(b), from which it is seen that six function blocks are required in the leapfrog model of Fig. 8.9.

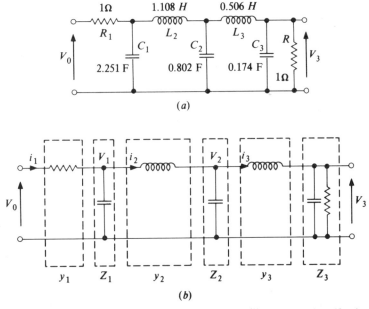

Fig. 8.8. Fifth-order lowpass Gaussian filter. (*a*) Passive filter; (*b*) immittance identification

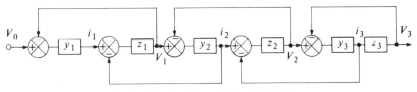

Fig. 8.9. Leapfrog model of passive ladder shown in Fig. 8.8(*a*), (*b*)

REALIZATION BY ANALOGUE SIMULATION

The analogue diagram may now be constructed by substitution of function blocks from Table 8.4. Note that y_1 does not contain an inductor and can thus be simulated by an inverting amplifier. However, since an inverter already exists in the first loop (as a sign changer), y_1, Z_1, and the first (upper) feedback loop can be replaced by a damped integrator of the form y_{1L}, but with $R_f = R$. Thus, in the final analogue diagram, shown in Fig. 8.10, i_1 is not available as an output variable. T_2', T_2, and T_3'' represent dual-input integrators, whilst sign

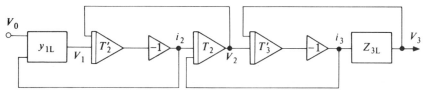

Fig. 8.10. Analogue diagram for fifth-order lowpass Gaussian filter

inversions are performed by two unity-gain VCVS's of the form described in Sec. 5.3.1 and positioned so that each feedback loop is of the correct sign. Hence, the total number of amplifiers employed is seven.

Since the terminating resistors are equal (and unity), the values of the capacitors in the analogue are identical to the corresponding reactive components of each immittance in the original ladder, i.e.,

$$C_1 = 2.251 \text{ F}, \quad C_2' = 1.108 \text{ F}, \quad C_2 = 0.802 \text{ F}$$
$$C_3' = 0.506 \text{ F} \quad \text{and} \quad C_3 = 0.174 \text{ F}$$

The design procedure can be developed for more general ladder structures encompassing elliptic and other complex filter functions, though additional

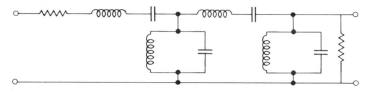

Fig. 8.11. Bandpass ladder transformed from lowpass prototype in Fig. 8.7

amplifiers may well be required.[16] The leapfrog method, though simulating transfer functions only, is economical in amplifiers, producing a regular pattern which is readily extended to higher-order systems (as exemplified in Ex. 8.2, Fig. 8.9) and which may be applied to other *LCR* ladders such as the transformed-bandpass network shown in Fig. 8.11.

8.2.2 *State-variable filter*

A state-variable[17] is one which defines the performance of a system and is subject to change. For example, the state of a circuit containing capacitive reactances can, at any given instant, be thought of as the voltage at a node. Formulation of the circuit differential equations in terms of 'state voltage' allows solution of output variables, given the input variables, with complete generality and the minimum number of independent system variables. The canonical form may be written as

$$\frac{d\bar{V}(t)}{dt} = \alpha \bar{V}(t) + \beta \bar{V}_{in}(t) \tag{8.7a}$$

and

$$\bar{V}_{out}(t) = \gamma \bar{V}(t) + \delta \bar{V}_{in}(t) \tag{8.7b}$$

where $\bar{V}$, $\bar{V}_{in}$, and $\bar{V}_{out}$ are the state, input, and output voltage vectors, respectively.

For the voltage-transfer function,

$$\frac{V_{out}}{V_{in}} = \frac{a_n s^n + a_{n-1} s^{n-1} + \cdots + a_1 s + a_0}{s^n + b_{n-1} s^{n-1} + \cdots + b_1 s + b_0} \tag{8.8}$$

Eqs. (8.7) represent[18]

$$\begin{bmatrix} \dot{V}_1 \\ \dot{V}_2 \\ \vdots \\ \dot{V}_{n-1} \\ \dot{V}_n \end{bmatrix} = \begin{bmatrix} 0 & 1 & \cdots & 0 & 0 \\ 0 & 0 & \cdots & 0 & 0 \\ \vdots & \vdots & & \vdots & \vdots \\ 0 & 0 & \cdots & 0 & 1 \\ -b_0 & -b_1 & \cdots & -b_{n-2} & -b_{n-1} \end{bmatrix} \begin{bmatrix} V_1 \\ V_2 \\ \vdots \\ V_{n-1} \\ V_n \end{bmatrix} + \begin{bmatrix} 0 \\ 0 \\ \vdots \\ 0 \\ 1 \end{bmatrix} V_{in} \tag{8.9a}$$

and

$$V_{out} = [a_0 \ a_1 \cdots a_{n-2} \ a_{n-1}] \begin{bmatrix} V_1 \\ V_2 \\ \vdots \\ V_{n-1} \\ V_n \end{bmatrix} + a_n V_{in} \tag{8.9b}$$

Converting into integral form, the state equations are associated with the general flowgraph (Appendix 4) of Fig. 8.12(a) and, in the time-invariant case, synthesis depends on the determination of the matrix elements. It is another multiple-loop system realizable by analogue methods.

Since roots of high-order polynomials are particularly sensitive to coefficients, nth order systems are customarily realized either as a cascade or a multiloop connection of second-order building blocks. Thus, the reduced (and simplified) biquadratic form shown in Fig. 8.12(b) will suffice for synthesis; the

REALIZATION BY ANALOGUE SIMULATION

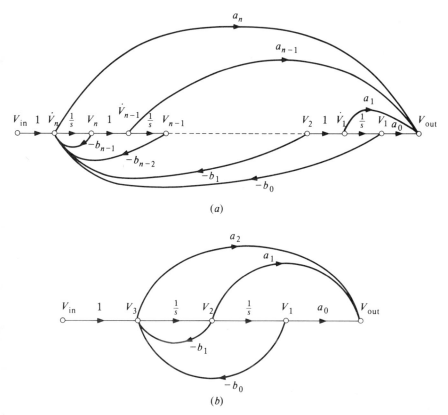

Fig. 8.12. State-variable systems. (a) Signal-flow graph for general state-variable system; (b) simplified signal-flow graph for second-order system

transfer function may be checked by Mason's rule (Appendix 4). Figure 8.12 is annotated for realization using non-inverting integrators (see Fig. 8.13(a)). However, it is advantageous from the point of view of passive sensitivity[19] and

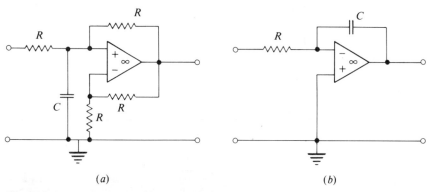

Fig. 8.13. Integrator circuits. (a) Non-inverting or positive integrator circuit; (b) inverting or negative integrator circuit

199

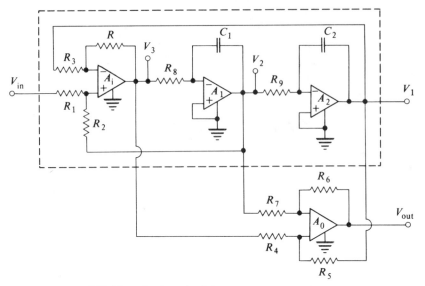

Fig. 8.14. State-variable filter using inverting integrators

component count to employ inverting integrators (Fig. 8.13(b)). Such implementation involves reversing the sign of alternate b coefficients in Fig. 8.12. The resulting electronic circuit is drawn[20] in Fig. 8.14. It is capable of realizing a variety of transfer functions according to the particular output connections used. This is emphasized by the inclusion of a dashed rectangle separating the output summer from the remainder of the network. Assuming $A_i = A_1 = A_2 = A_0 = \infty$, analysis yields the voltage transfer function as

$$\frac{V_{out}}{V_{in}} = \frac{R_2(R + R_3)R_5(R_6 + R_7)}{(R_1 + R_2)R_3(R_4 + R_5)R_7}$$

$$\times \left[\frac{C_1 C_2 R_8 R_9 s^2 + \left(\frac{[R_4 + R_5]R_6}{R_5[R_6 + R_7]}\right)R_9 C_2 s + \frac{R_4}{R_5}}{C_1 C_2 R_8 R_9 s^2 + \left(\frac{R_1[R + R_3]}{R_3[R_1 + R_2]}\right)R_9 C_2 s + \frac{R}{R_3}}\right] \quad (8.10)$$

For some transfer functions, the output summer is not required, namely:

(a) *the lowpass response* with

$$\frac{V_1}{V_{in}} = \left(\frac{R_2[R + R_3]}{R_3[R_1 + R_2]}\right)\bigg/ D(s) \quad (8.11a)$$

(b) *the bandpass response* with

$$\frac{V_2}{V_{in}} = -R_9 C_2 s \left(\frac{R_2[R + R_3]}{R_3[R_1 + R_2]}\right)\bigg/ D(s) \quad (8.11b)$$

where $D(s)$ is the denominator polynomial of Eq. (8.10).

REALIZATION BY ANALOGUE SIMULATION

These changes can be made by simple external connections to the universal active filter (UAF) module, modified forms of which are currently available from a number of manufacturers (Sec. 9.7).

Example 8.3 Design a bandpass filter having a Q factor of 25, a peak gain of 20 dB at 10 kHz, and denormalized to an impedance level of 10 kΩ.

SOLUTION   From Eqs. (8.10) and (8.11a),

$$\omega_0 = \sqrt{\left(\frac{R}{R_3}\right)\left(\frac{1}{C_1 C_2 R_8 R_9}\right)} \tag{8.12a}$$

$$Q = \frac{R_3(R_1 + R_2)}{R_1(R + R_3)} \sqrt{\left(\frac{R}{R_3}\right)\frac{C_1 R_8}{C_2 R_9}} \tag{8.12b}$$

and the peak or centreband gain is

$$K_0 = \left|\frac{V_2}{V_{in}}\right|_{max} = \frac{R_2}{R_1}. \tag{8.12c}$$

Since there are eight unknowns and only three equations, the designer has considerable freedom of choice. The peak gain is specified as 10 so that, if $R_1 = 1$, $R_2 = 10$. It is also convenient to let $C_1 = C_2 = 1$ and (without loss of generality) $\omega_0 = 1$, thereby reducing the design equations to

$$\omega_0^2 = \left(\frac{R}{R_3}\right)\left(\frac{1}{R_8 R_9}\right) = 1 \tag{8.13a}$$

and

$$Q = \frac{11 R_3}{(R + R_3)}\sqrt{\left(\frac{R}{R_3}\right)\left(\frac{R_8}{R_9}\right)} = 25 \tag{8.13b}$$

Further possibilities for equal-element allocation are restricted due to the form of Eqs. (8.13). One possibility is to let $R = R_3 = 1$ leading to $R_8 = 4.55$ Ω and $R_9 = 0.22$ Ω.

Denormalization, by the method in Appendix 1, results in the following element values:

$$C_1 = C_2 = \frac{1}{(2\pi \times 10^4)10^4} = 0.0016 \ \mu\text{F}$$

$$R = R_1 = R_3 = 10 \text{ k}\Omega, \quad R_2 = 100 \text{ k}\Omega,$$
$$R_8 = 45.5 \text{ k}\Omega, \quad \text{and} \quad R_9 = 2.20 \text{ k}\Omega.$$

Slight modification of the state-variable filter of Fig. 8.14 affords greater flexibility of design. If a resistor, R_{10}, is connected between ground and the non-inverting input of amplifier A_1, Eq. (8.12b) becomes

$$Q = \frac{R_3(R_1 + R_2 + R_1R_2/R_{10})}{R_1(R + R_3)}\sqrt{\left(\frac{R}{R_3}\right)\left(\frac{C_1R_8}{C_2R_9}\right)} \qquad (8.14)$$

while Eqs. (8.12a) and (8.12c) remain unaltered. Several elements may again be set equal as in Ex. 8.3. For example,

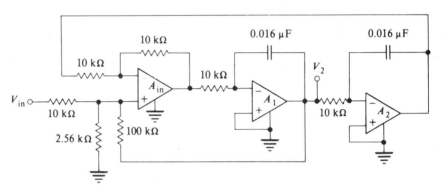

Fig. 8.15. State-variable bandpass filter having $Q = 25$, $f_0 = 10$ kHz, and midband gain of 20 dB

setting $R = R_3 = R_8 = R_9 = R_1 = C_1 = C_2 = 1$ and $R_2 = 10$ gives $\omega_0^2 = 1$, a midband gain of 20 dB and

$$Q = 5.5 + \frac{5}{R_{10}}$$

requiring $R_{10} = 0.256 \, \Omega$ for $Q = 25$. The final circuit, denormalized to 10 kHz and an impedance level of 10 kΩ, is shown in Fig. 8.15. It is suitable for simple tuning procedures since the gain and Q factor can be controlled independently while the centre frequency is determined by appropriate choice of R_8 and R_9 (assuming C_1 and C_2 to be fixed).

8.3 GIC methods

Attention is now turned to realizations utilizing the GIC of Fig. 5.23 which, by Table 5.5 and Eq. (5.37), has for the current-converter form the transmission matrix

$$[a] = \begin{bmatrix} 1 & 0 \\ 0 & Z_2Z_4/Z_1Z_3 \end{bmatrix} \qquad (8.15)$$

REALIZATION BY ANALOGUE SIMULATION

If the impedances $Z_1, \ldots, Z_4$ contain at least one capacitor and the remainder are resistors, Eq. (8.15) may be written

$$[a] = \begin{bmatrix} 1 & 0 \\ 0 & 1/k_1 s^n \end{bmatrix} \qquad (8.16)$$

where $n = \pm 1, \pm 2$.

In the realization procedures to be described, utilization of the GIC with $n = 1$ allows transformation of the inductive section of a network into a resistive array (Sec. 8.3.1). On the other hand, taking a GIC with $n = 2$ (but loaded at port 1, thereby making $n = -2$) transforms the entire network from LCR to DCR form, where D refers to the frequency-dependent negative resistor (FDNR). Finally, using a GIC with $n = 1$ or 2 allows direct implementation of a variety of second-order transfer functions (Sec. 8.3.3) with the additional advantage of being a grounded structure.

8.3.1 Simulation of inductor subnetworks

The system[21] in Fig. 8.16 consists of an array of n GIC's (sometimes termed positive immittance converters or PIC's) interconnected with an n-port resistive array. Using Table 5.5 and assuming identical conversion factors so that

$$k_1(s) = k_2(s) = \cdots k_n(s) = k(s)$$

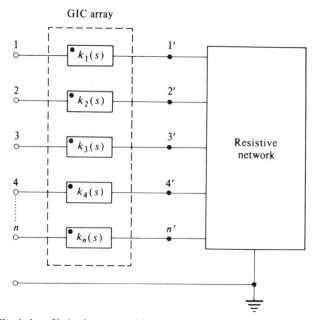

Fig. 8.16. Simulation of inductive or capacitive subnetwork

it is easily seen that

$$[V] = [V'] \tag{8.17a}$$

and

$$[I] = \frac{1}{k(s)}[I'] \tag{8.17b}$$

where $[V]$, $[V']$ and $[I]$, $[I']$ are column vectors representing voltages and currents at input and output ports of the GIC's. If $[Z]$ represents the impedance matrix of the total array and $[Z']$ represents the impedance matrix of the resistive network, Eqs. (8.17) may be rewritten as

$$[Z] = k(s)[Z'] \tag{8.18}$$

GIC's of the form shown in Fig. 5.23 have conversion factor

$$k(s) = \frac{Z_1 Z_3}{Z_2 Z_4}$$

and if Z_2 or Z_4 is a capacitor with the remaining three elements resistive, then $k(s) = sk$ and Eq. (8.18) becomes

$$[Z] = sk[Z'] \tag{8.19}$$

Hence, if $[Z']$ is a resistive matrix, $[Z]$ is an inductive matrix, i.e., a purely resistive array may be transformed (converted) into an inductive array having identical topology. The magnitude of each inductor is determined as

$$L_i = kR_i \tag{8.20}$$

where L_i and R_i represent elements in the same topological position. This is illustrated by the GIC-resistor connection in Fig. 8.17(a). The π-network has

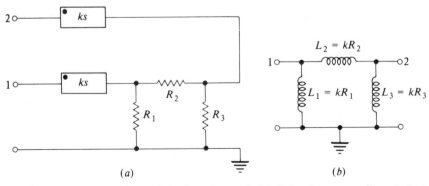

Fig. 8.17. Simulation of two-port inductive subnetwork. (a) GIC-resistor array; (b) equivalent inductive subnetwork

REALIZATION BY ANALOGUE SIMULATION

an impedance matrix given by

$$[Z'] = \begin{bmatrix} \dfrac{R_1(R_2 + R_3)}{(R_1 + R_2 + R_3)} & \dfrac{R_1 R_3}{(R_1 + R_2 + R_3)} \\ \dfrac{R_1 R_3}{(R_1 + R_2 + R_3)} & \dfrac{R_3(R_1 + R_2)}{(R_1 + R_2 + R_3)} \end{bmatrix}$$

However, the overall network has an impedance matrix determined by Eq. (8.19) and is equivalent to the inductive-π shown in Fig. 8.17(b). In this case, therefore, three inductors have been simulated using two GIC's, each of which includes one capacitor.

It is important to note that this method requires a number of GIC's (and hence capacitors) equal to the number of ports in the resistive array. Thus, if the array contains more internal elements than ports, fewer capacitors are required than would be the case in the direct simulation of inductance by capacitively-terminated gyrators. This, unfortunately, does not necessarily simplify the simulation of floating inductors, which must be regarded as two-ports.

The synthesis procedure may be summarized as follows:

(a) Obtain an LCR network to meet the design specifications.
(b) Isolate an n-port inductive subnetwork by extracting all inductors from the original structure.
(c) Form the GIC network (as in Fig. 8.16) having n ports.
(d) Replace the inductive subnetwork by a topologically equivalent resistive network with elements $R_i = L_i/k$, where k is the desired conversion factor (used here for convenience as the true conversion factor ks).

Example 8.4 Use the method of inductive subnetwork simulation to realize the fifth-order lowpass Gaussian filter of Ex. 8.2, having a 3 dB point at 1 kHz and denormalized to an impedance level of 600 Ω.

SOLUTION

(a) The normalized LCR filter is shown in Fig. 8.18(a).
(b) The inductive subnetwork is formed by separating the inductors from the remainder of the network, as shown in Fig. 8.18(b). In this case, a two-port subnetwork results.
(c) Three GIC's are required for the conversion.
(d) The inductive network may now be replaced by a resistive array and is connected, via the GIC's, to the remainder of the original filter as shown in Fig. 8.18(c) for the normalized realization with $k = 1$.

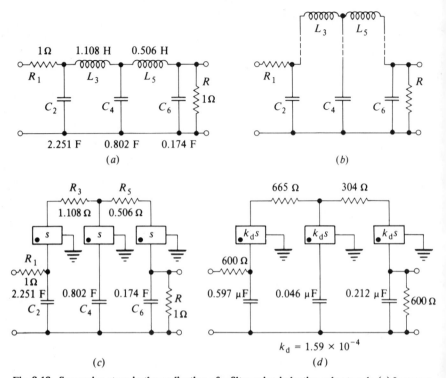

Fig. 8.18. Successive steps in the realization of a filter using inductive subnetwork. (a) Lowpass prototype; (b) separation of inductors from main network; (c) normalized network; (c) denormalized network

By the method in Appendix 1, for the specified impedance denormalization factor of 600 and the frequency denormalization factor ($2\pi \times 10^3$), the denormalized element values (R_d, C_d) are related to the normalized elements (R_n, C_n) as follows:

$$R_d = 600 R_n$$

$$C_d = \frac{C_n}{600 \times 2\pi \times 10^3}$$

The conversion factor, k, is unaffected by impedance denormalization since it is of the form

$$k = \frac{CR_a R_b}{R_c}$$

However, it is affected by frequency denormalization, so that the denormalized value of k is

$$k_d = \frac{k}{2\pi \times 10^3} = \frac{1}{2\pi \times 10^3} = 1.59 \times 10^{-4}$$

REALIZATION BY ANALOGUE SIMULATION

The final denormalized network is shown in Fig. 8.18(d) with the simulated values of inductance given by

$$L_3 = k_d R_3 = 1.59 \times 10^{-4} \times 665 = 105.8 \text{ mH}$$
$$L_5 = k_d R_5 = 1.59 \times 10^{-4} \times 304 = 48.4 \text{ mH}$$

These are the same values as those obtained if the inductors are denormalized directly from Fig. 8.18(a) using conventional techniques (Appendix 1). Hence, the GIC system is a true simulation of the denormalized ladder network.

An alternative approach to ladder realization is to use GIC's to simulate individual inductors, as described in Sec. 8.1. Of course, this necessitates the use of two GIC's to simulate each floating inductor and, for LCR simulation of moderate complexity, requires more active elements than the inductor subnetwork method.

8.3.2 FDNR synthesis

A simple technique[22] for the transformation of LCR structures is possible with the aid of frequency-dependent negative resistors (Sec. 5.4.4). Unlike the inductive subnetwork technique, in which the network is only partially transformed, the method to be described here is based on transformation of the entire network.

Since both voltage and current transfer functions of a network are ratios of polynomials in s, they are unaffected by scaling of the admittance matrix. Hence, if all admittances in a passive network are scaled by a factor s, the voltage transfer function remains unaltered and the topology is unchanged. However, transformation does occur in the network elements, as indicated below:

$$\begin{array}{llll}
\text{Admittance:} & Y(s) \Rightarrow sY(s): & \text{Transformed admittance} \\
\text{Capacitor:} & sC \Rightarrow s^2C: & \text{FDNR} \\
\text{Inductor:} & 1/sL \Rightarrow 1/L: & \text{Resistor} \\
\text{Resistor:} & 1/R \Rightarrow s/R: & \text{Capacitor}
\end{array}$$

The effect of such a transformation is thus to eliminate inductors and to create a new, but topologically equivalent, system which includes FDNR's. The simplicity of the method is illustrated in the following example.

Example 8.5 Use FDNR's to realize the fifth-order lowpass Gaussian filter of Ex. 8.4.

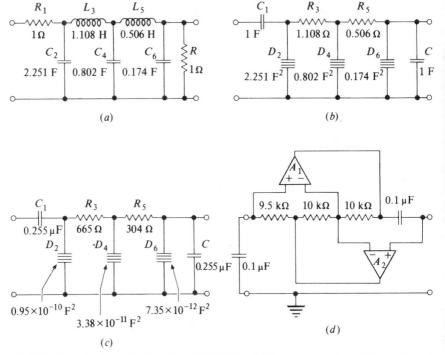

Fig. 8.19. Steps in the realization of an FDNR filter. (*a*) Lowpass prototype; (*b*) transformed (*DCR*) network; (*c*) normalized network; (*d*) realization of D_2 by capacitively terminated (at port 1) GIC

SOLUTION The now familiar passive prototype is shown in Fig. 8.19(*a*), application of the *s* transformation giving the *DCR* equivalent of Fig. 8.19(*b*). Treating the FDNR as a network element, the denormalized value of the constant is given by

$$D_d = \frac{D_n}{(\omega_n)^2 Z_n} \qquad (8.21)$$

D_n being the normalized value of the FDNR constant, ω_n and Z_n the frequency and impedance denormalizing factors, respectively. Again,

$$D_{2d} = \frac{2.251}{(2\pi \times 10^3)^2 \times 600} = 0.95 \times 10^{-10}$$

and the denormalized network is presented in Fig. 8.19(*c*). D_2, D_4, and D_6 may be realized by either resistively- or capacitively-terminated GIC's (Fig. 5.24). For example, the latter gives the realization

$$D_2 = kC' \qquad (8.22)$$

where C' is the value of terminating capacitance.

Using the impedance symbols Z_1, Z_2, Z_3, Z_4 for the internal elements of the GIC (Fig. 5.23), it is convenient to let Z_4 be a capacitor of value C', $Z_2 = Z_3 = R$, and $Z_1 = R'_1$. Then Eq. (8.22) gives $D_{2d} = R'_1 C'^2_1$.

Choosing $C' = 0.1$ µF gives $R_1 = 9.5$ kΩ. The realization of D_2 is shown in Fig. 8.19(d), in which $R = 10$ kΩ.

A disadvantage of the FDNR transformation is the creation of capacitors at input and output. This means there is no provision of input bias current for

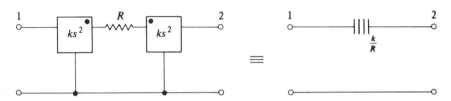

Fig. 8.20. Realization of a floating FDNR

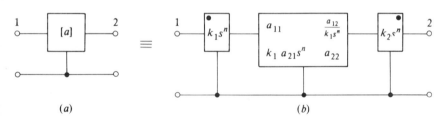

Fig. 8.21. Embedded network equivalence. (a) Original network; (b) embedded network

amplifier A_2 (Fig. 8.19(d)) and necessitates inclusion of resistive shunts. The presence of floating capacitors in a prototype network creates realization difficulties since the transformed system will contain floating FDNR's, each of which requires a pair of GIC's for simulation (Fig. 8.20). An alternative approach[23, 24] is to transform only part of the system and then embed this transformed subsection between two GIC's. Thus, if Fig. 8.21(a) represents the section to be transformed, its embedded equivalent is shown in Fig. 8.21(b), the composite matrix being given by

$$[a'] = \begin{bmatrix} 1 & 1 \\ 0 & 1/k_1 s^n \end{bmatrix} \begin{bmatrix} a_{11} & a_{12}/k_1 s^n \\ k_1 a_{21} s^n & a_{22} \end{bmatrix} \begin{bmatrix} 1 & 0 \\ 0 & k_2 s^n \end{bmatrix} = \begin{bmatrix} a_{11} & a_{12} k_2/k_1 \\ a_{21} & a_{22} k_2/k_1 \end{bmatrix}$$

This is equal to $[a]$ if $k_2 = k_1$. It is clear that this technique may also be used to preserve network input and output resistances, for matching purposes. In addition, it can convert a floating resistor to a floating inductor.

Example 8.6 Design a six-pole Butterworth bandpass filter using the embedded-FDNR approach.

SOLUTION The normalized prototype is shown in Fig. 8.22(a). If the total network is transformed (as in Ex. 8.5), two floating FDNR's result and the realization will require five GIC's, as shown in Fig. 8.22(b).

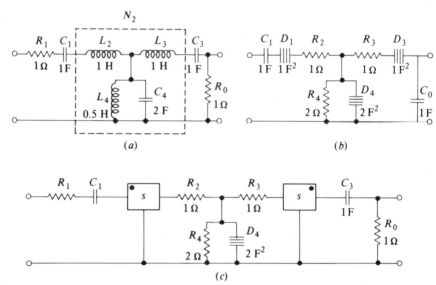

Fig. 8.22. Realization of bandpass filter. (a) Bandpass prototype; (b) realization using floating FDNR's; (c) embedded FDNR realization

The floating FDNR's can be avoided if network N_2 (Fig. 8.22(a)) is embedded between two GIC's. The FDNR which results is grounded and may be realized as described earlier in this section. The final, normalized, system is shown in Fig. 8.22(c). In each case, the GIC conversion factor is chosen as s^n, where $n = 1$ or 2.

8.3.3 GIC-derived structure

The circuit shown in Fig. 8.23 can be used to realize a variety of second-order sections such as lowpass, bandpass, and notch. Furthermore, by connecting[25] terminal 4 as the output of one stage via a resistor to terminal 3 of a succeeding stage, they may be cascaded for higher orders with the same facility as configurations in Chapters 6 and 7. Nodal analysis of Fig. 8.23 yields the following set of equations:

$$\begin{aligned}
\text{node 3} \quad & V_1 Y_5 - V_3(Y_1 + Y_5 + Y_6) + V_4 Y_1 = 0 \\
\text{node 5} \quad & V_4 Y_2 - V_5(Y_2 + Y_3) + V_6 Y_3 = 0 \\
\text{node 2} \quad & V_1 Y_7 - V_2(Y_4 + Y_7 + Y_8) + V_6 Y_4 = 0
\end{aligned}$$

REALIZATION BY ANALOGUE SIMULATION

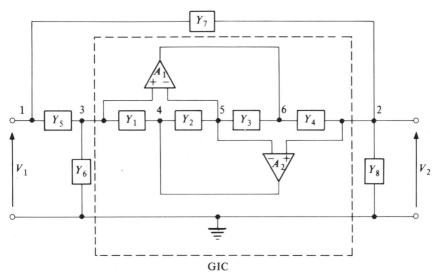

Fig. 8.23. GIC-derived structure

while, for the amplifiers,

$$V_6 = (V_3 - V_5)A_1$$

and

$$V_4 = (V_2 - V_5)A_2$$

The equations may be solved to yield the voltage transfer ratio:

$$\frac{V_4}{V_1} = \frac{Y_5 + k(s)(Y_7[1 + Y_6/Y_1] - Y_5 Y_8/Y_1)}{Y_5 + Y_6 + k(s)(Y_7 + Y_8)} \qquad (8.23)$$

If Y_3 is capacitive, $Y_8 = (G_8 + sC_8)$, $Y_5 = 0$, and the remaining elements are resistive, Eq. (8.24) yields the bandpass function

$$\frac{V_4}{V_1} = \frac{sC_3 G_7 (G_1 + G_6)}{s^2 C_3 C_8 G_1 + sC_3 G_1 (G_7 + G_8) + G_2 G_4 G_6}$$

Omitting G_8, the useful canonic realization with $C_3 = C_8 = C$, $G_1 = G_2 = G_4 = G_6 = G$ and $G_7 = G/Q$ emerges allowing independent Q control by a single resistor. Other second-order functions may be obtained by different choice of admittances and by taking the output from alternative nodes 2 and 6. For example, with output measured at node 6 and Y_1 capacitive, $Y_3 = (G_3 + sC_3)$, $Y_6 = Y_7 = 0$, and the remainder resistive, a lowpass function results.

8.4 Optimum design and sensitivity

The techniques discussed in the previous sections are summarized in Fig. 8.24. As well as providing a global picture of possible approaches, the diagram includes information on the number of active devices per simulated inductor and can thus serve as an initial design guide.

With the exception of the two-integrator-loop state-variable filter and the GIC-derived structure, the design methods have been based on the simulation of a double-terminated, lossless reactive ladder. The advantages of such an approach include both the availability of information associated with LC filters and their inherently low sensitivities. For these techniques, therefore, the question of optimum design is somewhat different from that discussed earlier. In Chapters 6 and 7, designs based on the same transfer function produced significant differences in sensitivity. By contrast, the ladder simulation techniques, whether direct or indirect, have a narrower spread in sensitivity because they are founded on the same basic structure. Consequently, for direct methods, optimum design will depend upon the quality of the simulated inductor and hence upon the active device whilst, for indirect realizations, the flexibility of approach constitutes an additional factor. For instance, the usefulness of the leapfrog method is dependent on the precise form of the ladder prototype. Similarly, techniques utilizing the GIC rely upon the exact form of the passive filter to be simulated.

Decisions on suitability of approach must also be taken with care. Thus, if a ladder prototype contains both floating and grounded coils, a detailed study may be necessary since gyrator simulation of

(*a*) each grounded coil requires one gyrator
(*b*) each floating coil requires two grounded gyrators or one semi-floating gyrator

while FDNR simulation of

(*a*) each grounded capacitor requires one FDNR
(*b*) each floating capacitor requires two grounded FDNR's or one floating FDNR.

Also, the number of active devices is clearly dependent upon the precise form of the passive model.

This chapter differs from Chapters 6 and 7 in one other and important respect. Whereas previous designs were based exclusively on second-order structures, ladder simulation is not so constrained. Hence, a comprehensive comparison based on biquadratic realizations would not be so meaningful. Furthermore, some of the methods (e.g., inductor-subnetwork simulation) would normally be used only for higher-order sections, and while Q and ω_0

REALIZATION BY ANALOGUE SIMULATION

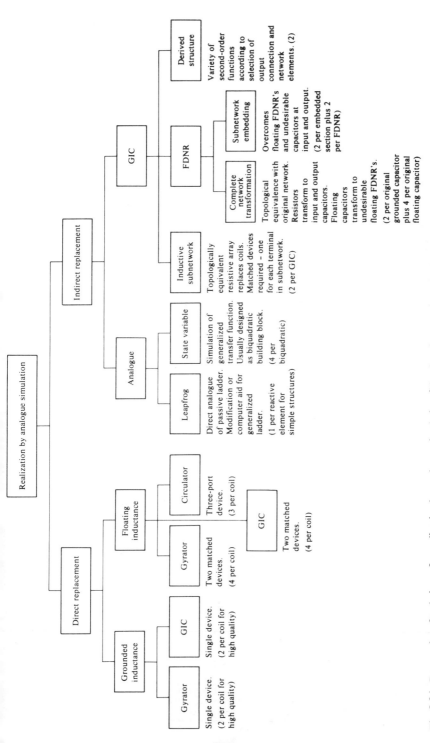

Fig. 8.24. Perspective of techniques for realization by analogue simulation. (Numbers in parentheses indicate basic number of amplifiers required.)

Table 8.5 Performance comparison for

Network structure	Circuit diagram	Performance	
		Q	ω_0
Passive		$\dfrac{G_0\sqrt{L_1/C_1}}{1+G_0/G_1}$	$\sqrt{\dfrac{1}{L_1C_1}}$
Gyrator realization (Fig. 8.2)		$\dfrac{\sqrt{G_aG_bC/C_1}}{\{G_AG_B/G_1+G_aG_b/G_0\}}$	$\sqrt{\dfrac{G_aG_b}{C_1C}}$
Circulator realization (Fig. 8.4)		$\dfrac{\sqrt{CG_0/C_1G}}{(1+G_0/G_1)}$	$\sqrt{\dfrac{GG_0}{CC_1}}$
State-variable (Fig. 8.14)		$\left(\dfrac{1+G_1/G_2}{1+G_3/G}\right)\sqrt{\dfrac{G_3G_9C_1}{GG_8C_2}}$	$\sqrt{\dfrac{G_3G_8G_9}{GC_1C_2}}$
Inductive subnetwork (GIC) (Fig. 8.17)		$\dfrac{G_0\sqrt{k_2/CG_2}}{\{1+k_2G_0/k_1G_1\}}$	$\sqrt{\dfrac{G_2}{k_2C}}$
FDNR (Fig. 8.20)		$\dfrac{C_1\sqrt{1/k_1G_2G_3}}{(1+k_2C_1/k_1C_2)}$	$\sqrt{\dfrac{G_2}{k_1G_3}}$
GIC-derived structure (Fig. 8.23)		$\dfrac{1}{G_7}\sqrt{\dfrac{C_8G_2G_4G_6}{C_3G_1}}$	$\sqrt{\dfrac{G_2G_4G_6}{C_3C_8G_1}}$

Note: ζ denotes active parameter.

second-order inductor replacement in RC-active filters

parameters for bandpass transfer function

ζ	$S^{Q_k}_\zeta$	$S^{\omega_0}_\zeta$	$|S^Q_{RC}|_{max}$	$|S^{\omega_0}_{RC}|_{max}$	Maximum component spread	Design Conditions
—	—	—	0.5	0.5	1	$G_0 = G_1 = 1$ $L = 2Q$ $C = 1/2Q$
G_a	0	0.5	0.5	0.5	1	$G_0 = G_1 = 1$ $C = C_1 = 1/2Q$ $G_a = G_b = G_A = G_B = 1/2Q$
G_b	0	0.5				
G_A	0.5	0				
G_B	0.5	0				
G	−0.5	0.5	0.5	0.5	1	$G_0 = G_1 = 1$ $C = C_1 = 1/2Q$ $G = 1/4Q^2$
A_1	$-Q/A_1$	$-1/2A_1$	1	0.5	$2Q$	$C_1 = C_2$ $G = G_1 = G_3 = G_8$ $= G_9 = (2Q-1)G_2$
A_2	$-Q/A_2$	$-1/2A_2$				
A_I	$1/A_I$	$-1/A_I$				
k_1	0.5	0	0.5	0.5	$2Q$	$G_0 = G_1 = 2QG_2 = 1$ $C = 1/2Q$ $k_1 = k_2 = 1$
k_2	0	−0.5				
k_1	0	−0.5	0.5	0.5	1	$C_1 = C_2 = 1$ $G_2 = G_3 = 1/2Q$ $k_1 = k_2 = 1$
k_2	−0.5	0				
A_1	$2Q/A_1$	$-(Q+1)/QA_1$	0.5	0.5	Q	$C_3 = C_8$ $G_1 = G_2 = G_4$ $= G_6 = QG_7$
A_2	$2Q/A_2$	$1/A_2$				

sensitivities exist, they are not so readily derived by analysis. Numerical evaluation of these for suitable responses is possible, but comparisons are more usually based on a study of the frequency-dependent classical sensitivity of the transfer function. This again requires the use of a digital computer.[26] The major disadvantage of the technique is that it produces no distinct figure of merit. However, especially if associated with a graphic display terminal,[27] the designer can readily observe the effects of adjusting components by prescribed amounts. This is akin to the insight provided to so many problems by an analogue computer. An alternative[28] is to use statistical measures (see Chapter 4) to predict the effects of component tolerances, drift and other parameters. It is usually sufficient to confine higher-order investigations to the passband and cut-off region, especially when dealing with bandpass filters.

To afford some degree of comparison, consider a second-order bandpass function realized by each of the methods listed in Table 8.5 which summarizes element values and performance parameters. Component spreads in each case are dependent on the precise form of the active element (gyrator, circulator or GIC) employed except for the state-variable structure which has a spread of $(2Q - 1)$ and the GIC-derived structure which has a spread of Q. Wherever possible, element values have been chosen equal.

Comparison of the sensitivity parameters must be approached with care since, as for component spread, the figures depend upon the exact form of the active device (except for the two exceptions noted above). For this reason, sensitivity values are presented with respect to a parameter ζ, a function of amplifier gain whose exact form can be derived by analysis of the particular gyrator or GIC employed in the design. The problem is more complex in the case of the circulator as is discussed later. As an example, with equal resistor values, the Antoniou gyrator of Fig. 5.22(c) has a forward gyration conductance G_A (see Eqs. (8.4) and (5.35)) given by

$$G_A = \frac{2 + A_1 + A_2 + A_1 A_2}{RA_1(2 + A_2)} \qquad (8.24)$$

from which

$$S^{G_A}_{A_1} = -\left(\frac{1}{1 + A_1}\right) \quad \text{and} \quad S^{G_A}_{A_2} = \left(\frac{1}{1 + A_2}\right)$$

Using Table 4.2,

$$S^{Q}_{A_i} = S^{Q}_{\zeta} S^{\zeta}_{A_i}$$

Hence, in this case,

$$S^{Q_k}_{A_1} = S^{Q_k}_{G_A} S^{G_A}_{A_1} = \frac{-0.5}{(1 + A_1)}$$

and

$$S_{A_2}^{Q_k} = S_{G_A}^{Q_k} S_{A_2}^{G_A} = \frac{0.5}{(1 + A_1)}$$

Similar calculations can be performed to obtain a complete set of sensitivity parameters for the system. The GIC, being derived from the Antoniou gyrator, may be treated in a similar fashion and is left as an exercise for the reader.

Analysis of the circulator is more awkward since the three-port device has an ideal admittance matrix of the form

$$[Y] = G \begin{bmatrix} 0 & 1 & -1 \\ -1 & 0 & 1 \\ 1 & -1 & 0 \end{bmatrix}$$

If the unity-valued elements are not exactly equal, parasitic elements arise in addition to the desired floating inductor. These elements can be critical to the passband-edge sensitivity of high-order filters.

The state-variable filter may be analysed in a straightforward manner though, due to the presence of three amplifiers, the resulting expressions are lengthy. The denominator coefficients (with the notation of Eq. (4.6)) may be written as

$$b_0 = 1 + \frac{1}{QA_2} + \frac{1}{A_1 A_2}\left(1 + \frac{2}{A_i}\right) \qquad (8.25a)$$

$$b_1 = \frac{1}{Q}\left(1 + \frac{1}{A_2}\right) + \left(1 + \frac{2}{A_i}\right)\left(\frac{1}{A_1}\left[1 + \frac{1}{A_2}\right] + \frac{1}{A_2}\left[1 + \frac{1}{A_1}\right]\right) \qquad (8.25b)$$

$$b_2 = \left(1 + \frac{1}{A_1}\right)\left(1 + \frac{1}{A_2}\right)\left(1 + \frac{2}{A_i}\right) \qquad (8.25c)$$

the elements having the values listed in Table 8.5 and the sensitivity expressions being evaluated on the basis of Eqs. (8.25(a)–(c)). A major feature of the state-variable filter is that it can be produced as a standard off-the-shelf item with the subsequent capability of realizing a wide range of transfer functions. In addition, for resonator sections, the capacitors and all but one of the resistors can be made equal.

For the GIC-derived structure, $S_{A_{1,2}}^Q$ is minimized by choosing[22]

$$(G_5 + G_6) = G_2 \quad \text{and} \quad \frac{(G_7 + G_8)}{G_4} = \frac{1}{Q}$$

while $S^{\omega_0}_{A_{1,2}}$ is minimized when

$$G_4(G_5 + G_6) = G_2(G_4 + G_7 + G_8)$$

Taking account of the finite amplifier gains and the element values listed in Table 8.5, the following expressions can be derived:

$$Q_k = \frac{\sqrt{\left[1 + \frac{2}{A_1}\left(1 + \frac{1}{A_2}\right)\left(1 + \frac{1}{Q}\right)\right]\left[1 + \frac{2}{A_2}\left(1 + \frac{1}{A_1}\right)\right]}}{\left\{\frac{1}{Q} + 2\left(1 + \frac{1}{Q}\right)\left(\frac{1}{A_2}\right)\left(1 + \frac{1}{A_1}\right) + \frac{2}{A_1}\left(1 + \frac{1}{A_2}\right)\right\}} \quad (8.26a)$$

and

$$\omega_0 = \sqrt{\frac{1 + \frac{2}{A_1}\left(1 + \frac{1}{A_2}\right)\left(1 + \frac{1}{Q}\right)}{1 + \frac{2}{A_2}\left(1 + \frac{1}{A_1}\right)}} \quad (8.26b)$$

and hence the sensitivity formulae evaluated for individual amplifiers.

For higher-order studies, the complexity and lack of definitive figures of merit necessitate the exercise of care when comparing different methods. In general, however, it has been found that direct-replacement methods and embedded techniques produce roughly the same sensitivity performance[24] while the leapfrog filter compares favourably with simulated inductance designs.

References

1. M. E. Van Valkenburg, *Introduction to Modern Network Synthesis*, Wiley, New York, 1960.
2. J. K. Skwirzynski, *Design Theory and Data for Electrical Filters*, Van Nostrand, London, 1965.
3. T. N. Rao, P. Gary, and R. W. Newcomb, 'Equivalent inductance and Q of a capacitor-loaded gyrator', *IEEE Journal*, **SC-2**, 32–33, 1967.
4. H. J Orchard, 'Inductorless filters', *Electron. Letters*, **2** (6), 224–225, 1966.
5. A. Antoniou, 'New gyrator circuits obtained by using nullors', *Electron. Letters*, **4** (5), 87–88, 1968.
6. H. J. Orchard and D. F. Sheahan, 'Inductorless bandpass filters', *IEEE Journal*, **SC-5**, 108–118, 1970.
7. D. F. Sheahan, 'Gyrator-flotation circuit', *Electron. Letters*, **3** (1), 39–40, 1967.
8. H. R. Trimmel and W. E. Heinlein, 'Fully floating chain-type gyrator circuit using operational transconductance amplifiers', *IEEE Trans.*, **CT-18**, 719–721, 1971.

9. W. E. Heinlein and W. H. Holmes, *Active Filters for Integrated Circuits*, Prentice-Hall, Englewood Cliffs, N.J., 1974, p. 297.
10. A. G. J. Holt and J. R. Taylor, 'Method of replacing ungrounded inductances by grounded gyrators', *Electron. Letters*, **1** (4), 105, 1965.
11. J. M. Rollett, 'Circulator-capacitor networks', *Electron. Letters*, **4** (26), 599–601, 1968.
12. C. A. Wass, *An Introduction to Electronic Analogue Computers*, Pergamon Press, London, 1955, pp. 93–94.
13. F. E. Girling and E. F. Good, 'Active filters', *Wireless World*, **76**, 341–345 and 445–450, 1970. The leapfrog method was first described considerably earlier by the same authors in RRE Memo No. 1177, Sept. 1955, being one of several far-sighted contributions by Messrs Girling and Good.
14. P. O. Brackett and A. S. Sedra, 'Active compensation for high-frequency effects in op-amp circuits with applications to active *RC* filters', *IEEE Trans.*, **CAS-23** (2), 68–72, 1976.
15. P. R. Geffe, *Simplified Modern Filter Design*, Iliffe, London and Rider, New York, 1963.
16. P. O. Brackett and A. S. Sedra, 'Direct SFG simulation of LC ladder networks with applications to active filter design', *IEEE Trans.*, **CAS-23**, 61–67, 1976.
17. R. W. Newcomb, *Active Integrated Circuit Synthesis*, Prentice-Hall, Englewood Cliffs, N.J., 1968, Chap. 3.
18. R. E. Kalman, 'Mathematical Design of Linear Dynamical Systems', *SIAM Journal*, Series A; *Control*, **1** (2), pp. 152–192, 1963.
19. L. C. Thomas, 'The biquad: Part 1—Some practical design considerations', *IEEE Trans.*, **CT-18** (3) 350–357, 1971.
20. W. J. Kerwin, L. P. Huelsman, and R. W. Newcomb, 'State-variable synthesis for insensitive integrated circuit transfer functions', *IEEE Journal*, **SC-2,** 87–92, 1967.
21. J. Gorski-Popiel, '*RC*-active synthesis using positive-immittance convertors', *Electron. Letters*, **3** (8), 381–382, 1967.
22. L. T. Bruton, 'Network transfer functions using the concept of frequency-dependent negative resistance', *IEEE Trans.*, **CT-16**, 406–408, 1969.
23. A. W. Keen and J. L. Glover, 'Active *RC* equivalents of *RCL* networks by similarity transformation', *Electron. Letters*, **7** (11), 288–290, 1971.
24. L. T. Bruton and A. B. Haase, 'Sensitivity of generalised immittance converter-embedded ladder structures', *IEEE Trans.*, **CAS-21** (2), 245–250, 1974.
25. B. B. Bhattacharyya, W. B. Mikhael, and A. Antoniou, 'Design of *RC*-active networks using generalized-immittance converters', *J. Franklin Inst.* **297** (1), 45–48, 1974.
26. D. A. Calahan, *Computer-aided Network Design*, McGraw-Hill, New York, 1971.
27. R. Spence and M. Apperley, 'The interactive-graphic man-computer dialogue in computer-aided circuit design', *IEEE Trans.*, **CAS-24** (2), 49–61, 1977.
28. 'Statistical circuit design', *Bell Systems Tech. J.*, **50** (4), pp. 1099–1310, 1971 (several papers).

9. Practical considerations

Earlier chapters have taken little account of the nature of real electrical components used in active-filter implementation. The particular technology employed to fabricate each network element influences its actual characteristics and therefore the overall filter performance. In order to understand these effects, it is necessary first to appreciate the technological processes described in Sec. 9.1 of this chapter. This is especially true of microelectronic realizations, for which the associated device imperfections should be included in network analysis if not synthesis. Also, in most practical situations, it is obviously advisable to consider the effects of changes in environmental conditions which have hitherto been assumed to be stable.

In view of the above discussion, the reader will now require a more global comparison to augment the passing remarks made in Chapters 6, 7, and 8 on related active-network families. The subject of tuning also merits attention from a more generalized foundation, as distinct to merely noting independent Q and ω_0 control for specific circuits, taking due cognizance of practical factors involved. Previous conditions have also been confined, in the main, to second-order sections whereas, in a real filtering problem (see Chapter 2), high orders are necessary to approximate a specified response. Consequently, Sec. 9.6 discusses the various philosophies of high-order design. However, the prime aim of the present chapter is to give an awareness of the practical outlet as well as the theoretical base. The preceding chapters have been largely divorced from commercial reality and it is necessary now to indicate which technologies and filter techniques are most appropriate for given tasks, stressing throughout the need for design compromise.

9.1 Component fabrication technologies

Recent years have seen significant improvements in the manufacturing methods and characteristics of discrete electronic components. While these extend the performance capability of active filters, the increasing complexity of circuitry in advanced communications systems has necessitated resort to more and more microminiaturized techniques. The most convenient form of the latter has been found to be *integrated circuits* (IC's), i.e., combinations of

PRACTICAL CONSIDERATIONS

interconnected elements inseparably associated on or within a continuous substrate.[1] These may be thought of as comprising three types of structure: monolithic, film, and hybrid combinations of the two made compatible by beam-lead sealed-junction methods. In all cases, there is a reduction in size, improvement in reliability owing to the elimination of wiring, and, most importantly, a significant decrease in large-scale production costs. Indeed, full exploitation of the possibilities afforded by active filters has only become possible through the availability of mass-production methods based on microelectronic technology. By their very nature, components so fabricated are well correlated owing to their dependence on a small number of process variables and have close tracking properties, but parasitic effects tend to be intensified by microminiaturization. The significance of these characteristics is discussed in various sections of this chapter.

9.1.1 Monolithic integrated circuits

Since the advent of the discrete junction transistor, progress in semiconductor planar process technology has enabled formation of entire circuits on a single monolith or chip (i.e., a compound of heterogeneous solids governed by diffusion laws). In this manner, highly sophisticated microelectronic circuits can be fabricated[2] by means of a series of selective dopant diffusions into a silicon substrate, as illustrated on the left-hand side of Fig. 9.1. The most important of these circuit devices in filter design is undoubtedly the operational amplifier (OA) whose fundamental properties have been discussed in Sec. 5.2. As first introduced, this required external compensation (see Sec. 9.2) but later designs incorporate internal compensation as witnessed by the much-used μA741. More recent innovations have allowed integral inclusions of field-effect processes leading to the bi-FET OA available in quad form as the TL084 and

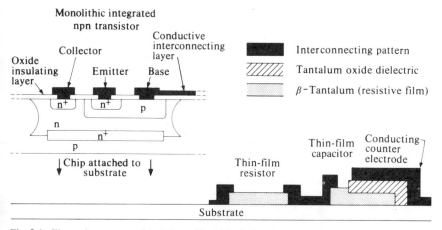

Fig. 9.1. Illustrative cross-sectional view of hybrid thin-film integrated circuit

Table 9.1 Typical performance parameters for amplifiers used in active filters

Parameter	Symbol	Amplifier type				
		μA741	LM301	LM310	L144	LH0062
Open-loop gain	A_0	2×10^5	1.6×10^5	0.9999	3×10^4	1.6×10^5
Gain-bandwidth product (MHz)	ω_t	1	6*	10	0.6	15
Output resistance (Ω)	R_{out}	75	—	0.75	—	75
Input resistance (MΩ)	R_{in}	0.5	2	10^6	1	10^6
Input capacitance (pF)	C_{in}	1.4	—	1.5	—	4
Slew rate (V μs^{-1})	ρ	0.5	10	30	0.4	70
Common-mode rejection ratio (dB)	CMRR	90	90	—	80	90
Input offset voltage (mV)	V_{IO}	1	2	2.5	1	2
Input offset current (μA)	I_{IO}	0.2	0.003	0.002	0.05	10^{-6}
Bias current (μA)	I_{IB}	0.6	0.07	0.01	0.13	10^{-5}
Device dissipation (mW)	P_D	50	60	120	15 (per OA)	200
Offset voltage temperature coefficient (μV K^{-1})	TCV	6	6	10	3	10
Remarks		Dual (μA747) and quad (TL084) versions available	External compensation required. *3 pF compensation	Voltage follower. Compensation required for symmetrical o/p current swing	Triple OA Low power	High quality. Extremely low bias currents

222

Table 9.2 Typical characteristics for some resistor types used in active filters

Property	Discrete			Thick film	Thin film		Diffused
	Carbon film	Metal film	Wire wound	Cermet	Nichrome	Tantalum nitride	Silicon
Resistance range	$10\,\Omega \rightarrow 2\,M\Omega$	$10\,\Omega \rightarrow 1\,M\Omega$	$10\,\Omega \rightarrow 200\,k\Omega$	$10\,\Omega \rightarrow 10\,M\Omega$	$15\,\Omega \rightarrow 150\,k\Omega$	$10\,\Omega \rightarrow 1\,M\Omega$	$50\,\Omega \rightarrow 50\,k\Omega$
Resistance per square (Ω)	—	—	—	$10 \rightarrow 10^5$	$100 \rightarrow 300$	$25 \rightarrow 300$	$100 \rightarrow 400$
Best untrimmed tolerance (per cent)	1	0.1	0.05	15	0.5	0.1	5
Temperature coefficient (p.p.m. K^{-1})	± 250	± 25	± 10	± 250	± 100	± 200	$-50 \rightarrow +5000$
Tracking (p.p.m. K^{-1})	—	—	—	± 2.5	± 10	± 5	± 50
Ageing (per cent per 20 years)	$2 \rightarrow 5$	0.1	0.01	1	2.5	0.1	—

Table 9.3 Typical characteristics for some capacitor types used in active filters

Property	Capacitor type						
	Mica	NPO ceramic	Mylar	Polycarbonate	Polystyrene	Ta_2O_5 thin film	Diffused silicon
Capacitance range	100 pF → 10 nF	1 pF → 10 nF	1 nF → 10 µF	1 nF → 10 µF	1 nF → 0.1 µF	10 pF → 10 nF	<100 pF
Capacitance ($\mu F\ m^{-3}$)†	3.3×10^3	2×10^5	1.1×10^4	1.1×10^4	1.1×10^4	(10^3)‡	(1.5×10^3)‡
Q_c (at 1 kHz)	1000	1000	200	2000	4000	400	—
Best tolerance (per cent)	±1	±1	±1	±1	±1	±20	±25
Temperature coefficient (p.p.m. K^{-1})	+40	±30	+800	+120	−100	+200	+200
Ageing (per cent per 20 years)	0.1	0.05	2	—	0.2	0.5	—

† For a 10 nF capacitor.
‡ These two values are given in microfarads per *square* metre.

offering several improved characteristics. Typical performance parameters for a selection of OA's appropriate for use in active filters are listed in Table 9.1.

The major limitation of the monolithic semiconductor process for precise analogue applications is its inability to produce suitable passive components. Diffused resistors occupy excessive chip area, are of loose absolute tolerance, and cannot be adjusted in value, though they track well with temperature. Junction capacitors, while allowing voltage tuning, suffer from the same disadvantages and are only incorporated in OA's to provide internal compensation, never as frequency-determining elements which must be formed of another technology. The passive properties are given in the right-hand columns of Tables 9.2 and 9.3 for contrast with, rather than as alternatives to, the other types.

9.1.2 Discrete components

The traditional components are still employed by several filter manufacturers, especially for instrument and test filters (see Sec. 9.7), where they offer the merits of high precision, proven reliability, and a wide range of values. Table 9.2 presents a selection of resistor types,[3] metal film being the least expensive and therefore the usual choice for precision applications. The tabular entries are consensus figures abstracted from several sources and are thus subject to change with ever-advancing technology. This is most true in the case of capacitors, for which a considerable variety of dielectric materials is now available. Table 9.3 indicates the excellent performance of ceramic, mica, polycarbonate, and polystyrene, though the extremely low loss of the latter is somewhat offset by its limited range of values and its sensitivity to soldering heat at values below 5 nF.

9.1.3 Thin-film integrated components

These have already been introduced in Sec. 3.5 in the form of distributed-parameter structures. Thin-film integrated technology developed from discrete-component film processes and produces essentially the same electrical properties. The films are produced[4] in evacuated vessels by either evaporation and deposition or by high-voltage sputtering onto a flat substrate of alumina or glazed ceramic. The controlled film thickness is normally less than 1 μm and can be as low as 1 nm. Resistive and conductive patterns are then made by selective etching to form the required circuit array, a subsequent deposition of gold, aluminium, or other suitable conductor being necessary for the interconnecting paths.

Tantalum technology[5] is especially suited for fabrication of active filter components, resistors being formed by sputtered-b.c.c. (body-centred cubic) or β tantalum or by high-pressure-sputtered oxygen-doped tantalum nitride. After vacuum deposition of the conductive links, this results in stable precision resistors which, as seen from Table 9.2, are superior to the nichrome variety though the latter only require a vacuum deposition cycle. All types of thin-film integrated resistor can be increased in value by a variety of methods, the most common of which are anodization, diamond scribing, abrasive and laser trimming. Further extension of resistance range can be obtained[6] by resort to meandered track (see Plate 1), but the finite area ultimately proscribes too large a resistor ratio or total resistance in the circuit. Tantalum technology also lends itself to fabrication of capacitors by oxidation. This affords high dielectric strength in very thin films. Though an order better than alternative dielectrics such as silicon monoxide, substrate area again limits capacitance values along with production yield at the upper end of the range and tolerance at the lower end. The detailed construction of tantalum passive components is illustrated in cross section by Fig. 9.1. In this, as in other integrated-circuit constructions, it is convenient to work in terms of the quantities of

$$\text{resistance per square, } R = \frac{\text{total resistance}}{\text{aspect ratio}} = R_t \cdot \frac{W}{L} = Wr_0 \quad (9.1)$$

and

$$\text{capacitance per unit area, } C = \frac{\text{total capacitance}}{\text{surface area}} = \frac{C_t}{WL} = \frac{c_0}{W} \quad (9.2)$$

rather than the per-unit-length quantities of Sec. 3.5, where R_t = total resistance of section, C_t = total capacitance of section, W = width of section, L = length of section (as in Fig. 3.13), and L/W = aspect ratio.

9.1.4 Thick-film integrated components

Thick films are distinguished from thin ones by having thicknesses in excess of 1 μm as well as displaying characteristics determined more by the film material and its subsequent treatment than by the method of deposition. The associated technological process[4,7] is altogether less precise than that just described in Sec. 9.1.3. It consists of forming resistive patterns on an alumina–ceramic substrate by a screen-printing operation in which a ceramic–metal paste, or *cermet*, is selectively squeezed through a fine stainless-steel mesh. Following a heat-treatment cycle, the resulting resistors exhibit excellent electrical properties as may be judged *vis-à-vis* thin films in Table 9.2. Despite a somewhat wide initial tolerance, resistor values are readily trimmed (see Plate 2) within one per cent by similar methods to those mentioned above. The high resistance per square available obviates the necessity for meandering which is, in any case, undesirable from the point of view of hot spots and trimming. The shortcoming of thick-film technology lies in the difficulty of producing adequate screened capacitors. The modest capacitance per unit area and relatively large losses of these have so far restricted their application and usually preclude realization of distributed $\overline{RC}$ structures. However, the possibility of reduced costs and improved performance[8] may offer promise for the future.

9.1.5 Hybrid circuit technology

In order to take advantage of the better qualities of each type of component fabrication, they may be used in combination to complement each other as a *hybrid circuit*.[9] This usually entails bonding the active-element chips to the passive substrate, as illustrated in Fig. 9.1. Here, only a single diffused transistor is included, but there would normally, of course, be a complete integrated OA chip either (a) back-bonded to the substrate and subsequently wire-bonded to the circuit by ultrasonic or thermocompression means; or (b) face-bonded directly to the circuit pattern, as in the case of flip chips. Since capacitor values are severely limited by practical substrate area, low-frequency filters often require passive chip devices to be attached to the substrate in a similar fashion to the OA's. This is particularly so for thick-film technology in which multilayer chip capacitors formed by screening noble-metal electrodes on strips of unfired ceramic are attached to the substrate by methods such as the TAB process[10] to reduce assembly costs. However, all such bondings are of semi-discrete components and consequently do not correlate or track as well as integrated units. For economy of area, occasional use is made of both sides of the substrate, one for the film resistors and the other for bonded OA's and capacitors.

9.1.6 Production processes and trimming

Active filters achieve the benefits of large-scale manufacture by the deployment of automated production lines serving high-volume demands for standardized units such as the STAR circuit of Sec. 7.3.3 and the two-integrator-loop or state-variable system of Sec. 8.2.2. A key factor has been the development of computer-controlled trimming equipment which affords the only feasible tuning procedure (see Sec. 9.5) for substantial production quantities. Highest operating speed and most accurate control of cut is at present offered by laser trimming,[11] which essentially increases the value of individual resistors by vapourizing or burning away a narrow slot in the original resistor pattern, as illustrated for the Lim-type circuit in Plate 2. The adjusted resistor values can be monitored as part of a bridge circuit or, alternatively, the entire filter circuit can be brought within specification. In the latter case, the finite settling time to allow trim-induced transients to subside imposes a basic speed limitation[12] which must be taken into account when optimum trimming algorithms are being devised. The overriding aim in production is, of course, the attainment of maximum yield by proper use of quality control[13] at each stage of manufacture, thereby minimizing the unnecessary testing of substandard specimens.

There can be no universal answer to the choice of technology used to implement active filters in all circumstances. The economics depend on the quantities involved, the total number of identical (or, at least, isotopic) circuits, the number of components per package which require trimming together with the degree of adjustment necessary. Where high-precision elements and performance are at a premium, thin films are preferable to thick, though the simplicity, relative cheapness, and continuing advances[14] of the latter approach make it an attractive proposition for quantities in the range of 10 000 to 100 000. However, the decisions involved are intricate and several manufacturers utilize both film capabilities, occasionally mixing the technologies on a single substrate. By contrast, for production quantities up to several thousand units, discrete realizations frequently provide the optimum solution.

9.2 Device imperfections

Hitherto in this text, both active and passive elements of filter circuits have been considered to be ideal electrical devices. In practice, however, there are parasitic and other unwanted effects associated with the fabrication techniques just discussed. These must either be accommodated in design or eliminated by compensation, attention being focused mainly on the active devices as these have wider production spreads. While emitter-follower transistors can be used directly in unity-gain realizations, general active networks have been seen to require integrated operational amplifiers. As indicated in Chapter 5, the voltage operational amplifier, in particular, is the essence of most present-day active filters and its non-idealities are therefore of most concern. In addition, gyrator performance can be impaired as a result of component imperfections, giving rise to phase shift in gyration conductance.

9.2.1 The non-ideal operational amplifier

The admittance matrix characterization presented for the differential-input model in Table 5.2 readily incorporates practical imperfections arising from the structure of the operational amplifier. To this end, the equivalent circuit is redrawn in Fig. 9.2, illustrating the predominant linear non-idealities: namely,

the finite frequency-dependent differential open-loop gain $= A(s)$,

the single-ended output impedance $= Z_{out} = 1/Y_{44}$,

the differential-input impedance $= Z_{in} = 1/Y_{12}$,

and the common-mode input impedances $= Z_{in_{cm}} = 1/Y_{11} = 1/Y_{22}$.

PRACTICAL CONSIDERATIONS

They are here expressed for a symmetrical-input system in terms of the admittance parameters defined by Eq. (5.2). The common-mode rejection ratio (CMRR) = $A_{cm} = |A/A_c|$, where $A_c \ll A$ is the common-mode signal gain with respect to the average as distinct to the difference level and can normally be neglected. As in Sec. 5.2, the model disregards the influence of internal feedback as well as error signals emanating from the d.c. bias supplies and noise since these have little effect at kilohertz frequencies.

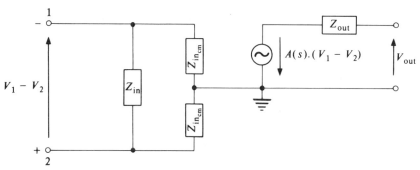

Fig. 9.2. Linear model of non-ideal operational amplifier

Finite amplifier gain is of significance in unconditionally stable systems since it severely limits the available Q factor as exemplified by Eq. (7.38) for the multiple-loop feedback circuit of Fig. 7.5. This emphasizes the need to restrict high-gain negative-feedback structures to moderate-Q applications, though utilization of the non-inverting OA capability in the differential-input single-feedback circuit of Fig. 7.6 relaxes the gain restriction by Eq. (7.22) to $A \gg 4Q$. Introduction of auxiliary positive feedback as in the modified MLFB circuit of Fig. 7.12 must not be carried to excess in improving Q for fear of inducing instability. It should be restricted to Q-enhancement factors less than 2:1 and resultant maximum selectivity of 60. The FEN system of Fig. 6.19 exhibits the most effective form of Q enhancement, increasing the passive selectivity by the straight multiplier of Eq. (6.59) rather than by a denominator subtraction. The transmission parameters of gyrators and GIC's are also altered by finite amplifier gain, the subsequent effect of perturbed conversion factors on the ladder realizations being readily investigated[15] by the incorporation of suitable models for simulated inductors or FDNR's and subsequent computer evaluation.

The frequency dependence of the voltage gain is characterized by the monotonic decrease in Fig. 9.3 with Bode-plot approximation indicated by dotted straight lines. The latter represent real poles giving cut-off rates of −6, −12, and −18 dB per octave at break frequencies ω_1, ω_2, and ω_3, respectively. A phase shift of 90° occurs between each pair of break frequencies. In any practical application, the amplifier forms part of a closed loop which is *strictly*

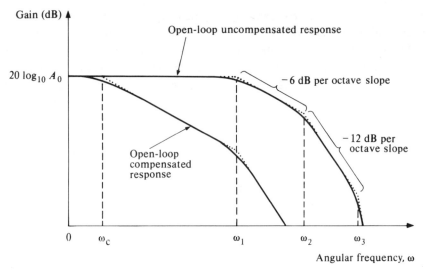

Fig. 9.3. Frequency response of open-loop voltage operational amplifier with internal compensation

stable (see Sec. 4.2.1) only if the phase of the loop gain is less than 180° at the frequency for which the magnitude is 0 dB. Otherwise, the negative-feedback circuit oscillates as the phase reversals reinforce. In order to ensure an adequate stability margin, it is necessary to impose a 6 dB-per-octave roll off at frequencies for which the loop gain >1. This is usually achieved by means of *internal compensation*, i.e., provision of additional passive elements (usually a capacitor C_c associated with resistance R_c integral with the amplifier) which effectively swamps the effect of other internal reactances (emanating from semiconductor junctions) and provides a dominant pole as the first break frequency ω_c of the open-loop gain

$$A(s) = \frac{A_o}{s/\omega_c + 1} = \frac{\omega_t}{s + \omega_c} \simeq \frac{\omega_t}{s} \qquad (9.3)$$

where A_o is the d.c. open-loop gain, $\omega_t = A_o \omega_c$ is the *gain-bandwidth product* (GBWP) with selected values given in Table 9.1, and $\omega_c = 1/R_c C_c < \omega_1$ is the open-loop compensated half-power corner frequency. Of course, as is evident from Fig. 9.3, this restricts the gain available at higher frequencies and thereby limits filter performance.

Considerations of amplifier gain-bandwidth product give yet another basis for comparison of active filter circuits. Replacement of A by Eq. (9.3) in the transfer ratios of earlier chapters leads to higher-order functions. The resulting shift in Q and centre frequency may be judged[16] by factorizing real poles from the denominator and examining the residual quadratic. For identical amplifiers, the cubic denominator function becomes

$$D(s) = b_3' \left(\frac{s}{\omega_0}\right)^3 + b_2' \left(\frac{s}{\omega_0}\right)^2 + b_1' \left(\frac{s}{\omega_0}\right) + 1$$

$$\triangleq \left(\frac{s}{\omega_f} + 1\right) \left[b_2' \left(\frac{s}{\omega_0}\right)^2 + (b_1' - b_3')\left(\frac{s}{\omega_0}\right) + 1\right] \quad (9.4)$$

provided $b_1' - 1$, $b_2' - 1$, and $b_3' \ll 1$. The actual selectivity Q_e and centre frequency ω_{0e} are simply found by application of Eq. (4.7). Alternatively, a perturbation technique based on substitutions of the type

$$\hat{\omega}_0 = \omega_0(1 + \delta\omega_0) \quad \text{and} \quad \hat{Q} = Q(1 + \delta Q) \quad (9.5)$$

can be employed[17] for algebraic convenience, allowing evaluation of single and multiple amplifiers to any degree of accuracy. $\delta\omega_0 = \Delta\omega_0/\omega_0$ and $\delta Q = \Delta Q/Q$ are, in fact, the *relative variations*. Application of the methods of Eq. (4.43) for high Q and identical amplifiers neglecting higher powers and crossproducts of $\delta\omega_0$ and δQ leads, after simultaneous solution,[18] to

$$\delta\omega_0 = \frac{n_0 - n_2}{2A_0} + \frac{n_2 - n_1}{2Q}\left(\frac{\omega_0}{\omega_t}\right) \quad (9.6)$$

and

$$\delta Q = \frac{n_0 + n_2 - 2n_1}{2A_0} + \left[(n_2 - n_0)Q + \frac{n_1 - n_2}{2Q}\right]\left(\frac{\omega_0}{\omega_t}\right) \quad (9.7)$$

The desirability of equalized remainder coefficients $n_0 = n_2$ is again apparent as in Eq. (4.44). Approximate results for a sample of networks in earlier chapters are presented in Table 9.4. With the exception of the multiple-loop-feedback circuit, the frequency perturbations are all negative and comparable, but there is much more variation in Q perturbations. The significance of this on high-frequency selective performance is to be discussed in Sec. 9.4.2. Notice in passing, the important identity

$$\frac{S_{\omega_t}^Q}{S_{\omega_t}^{\omega_0}} = \frac{\delta Q}{\delta \omega_0} = \frac{\Delta Q/Q}{\Delta \omega_0/\omega_0} = 2Q\left(\frac{\omega_t}{A_0\omega_0}\right) - 1 \quad (9.8)$$

which represents modification of Eq. (4.39) for the case of finite gain-bandwidth product. It is strictly applicable only under the condition $n_0 = n_2$, but is conditionally approached at high Q by circuits not of this family such as the GIC-derived structure. General opinion is that sensitivity to ω_t (which is related to group delay) is more crucial in filter operation than sensitivity to A_0

Table 9.4 Approximate frequency and Q perturbations with respect to equal gain-bandwidth product for a sample of second-order high-Q active networks

Network	$\delta\omega_0$	δQ
Sallen and Key single-source PFB (Fig. 6.2)	$-9\left(\dfrac{\omega_0}{\omega_t}\right)$	$9\left(\dfrac{\omega_0}{\omega_t}\right) - \dfrac{18Q}{A_0}$
Lim BP-PFB (Fig. 6.22)	$-2\left(\dfrac{\omega_0}{\omega_t}\right)$	$2\left(\dfrac{\omega_0}{\omega_t}\right) - \dfrac{4Q}{A_0}$
RC-embedded BP-PFB (Table 6.4)	$-4\left(\dfrac{\omega_0}{\omega_t}\right)$	$4\left(\dfrac{\omega_0}{\omega_t}\right) - \dfrac{8Q}{A_0}$
Multiple-loop feedback (MLFB) (Fig. 7.5)	$-Q\left(\dfrac{\omega_0}{\omega_t}\right)$	$Q\left(\dfrac{\omega_0}{\omega_t}\right) - \dfrac{2Q^2}{A_0}$
State-variable integrator loop (Fig. 8.14)	$-\left(\dfrac{\omega_0}{\omega_t}\right)$	$4Q\left(\dfrac{\omega_0}{\omega_t}\right) - \dfrac{2Q}{A_0}$
GIC derived (Fig. 8.23)	$-2\left(\dfrac{\omega_0}{\omega_t}\right)$	$\dfrac{1}{Q}\left(\dfrac{\omega_0}{\omega_t}\right) - \dfrac{4Q}{A_0}$
Allpass pair (Fig. 9.4)	$-\dfrac{5}{2}\left(\dfrac{\omega_0}{\omega_t}\right)$	$\dfrac{5}{2}\left(\dfrac{\omega_0}{\omega_t}\right) - \dfrac{5Q}{A_0}$

(which is merely related to attenuation), especially at frequencies above sub-audio.

Typical values of output impedance are listed in Table 9.1. This parameter is usually taken as being purely real, representing the terminal resistance of the

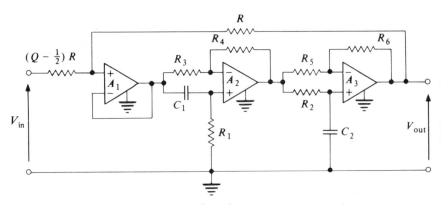

Fig. 9.4. Positive-feedback allpass pair configuration

emitter-follower output stage of the operational amplifier. Whereas R_{out} by itself has negligible effect on filter performance, there is a powerful interaction with gain-bandwidth product (see the representative responses of Fig. 9.5) which can disturb ω_t counterbalancing and, if R_{out} is large enough, provoke instability. The differential input impedance Z_{in} comprises a high (particularly for FET designs) resistance R_{in} in shunt with a small capacitance C_{in}. Often, $Z_{in_{cm}} \gg Z_{in}$ and for a single-sided input connection they degenerate into a single effective input impedance. As illustrated in Fig. 9.5, R_{in} considered as the

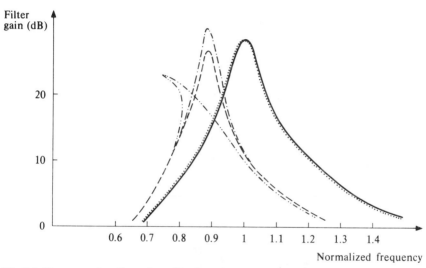

Fig. 9.5. Representative effects of amplifier imperfections on the frequency response of an MLFB bandpass filter. ———, Ideal; ------, with R_{out} and R_{in}; — — —, with ω_t; — · —, with R_{out}, R_{in}, and ω_t; — · · —, with linear and non-linear imperfections

sole imperfection also gives rise to a frequency response almost indistinguishable from the ideal. Appropriate choice of the relative resistance level of the passive sections can minimize the effects of parasitic impedance, this being of particular relevance in filter circuits with controlled feedback paths (e.g., Fig. 7.12).

9.2.2 Active-R filters

The external capacitors described in Sec. 9.1 are extravagant with chip area and difficult to trim. They may be eliminated by utilizing the otherwise troublesome internally compensated single-pole representation of practical operational amplifiers for the formation of transfer functions. The advantages of an all-resistive passive section in circuit integration are obvious and, at the same time, the frequency range of operation is naturally extended. The

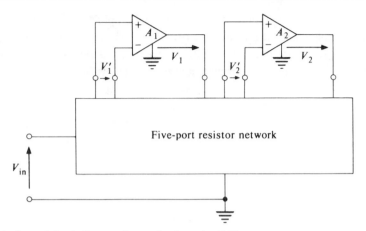

Fig. 9.6. General circuit diagram of second-order active-R filter

realization of a second-order filter requires[19] the embedding of two OA's in a five-port passive R network as in Fig. 9.6. By superposition,

$$[V'(s)] = [F][V(s)] + [f][V_{in}(s)] \tag{9.9}$$

where

$$[F] = \begin{bmatrix} F_{11} & F_{12} \\ F_{21} & F_{22} \end{bmatrix} \quad \text{and} \quad [f] = \begin{bmatrix} f_{1i} \\ f_{2i} \end{bmatrix}$$

By Eq. (9.3),

$$[V'(s)] = [A(s)]^{-1}[V(s)] = \begin{bmatrix} \omega_{t_1}/s & 0 \\ 0 & \omega_{t_2}/s \end{bmatrix}^{-1} \begin{bmatrix} V_1(s) \\ V_2(s) \end{bmatrix} \tag{9.10}$$

Hence, solving

$$[V(s)] = -[[F] - [A(s)]^{-1}]^{-1}[f][V_{in}(s)]$$

and, at the second amplifier,

$$\frac{V_2(s)}{V_{in}(s)} = \frac{\dfrac{1}{F_{11}F_{22} - F_{12}F_{21}}\left[\left(\dfrac{f_{2i}}{\omega_{t_1}}\right) + (F_{21}f_{1i} - F_{11}f_{2i})\right]}{\left(\dfrac{s}{\omega_0}\right)^2 + \dfrac{1}{Q}\left(\dfrac{s}{\omega_0}\right) + 1} \tag{9.11}$$

where

$$\omega_0 = \sqrt{F_{11}F_{22} - F_{12}F_{21}} \cdot \sqrt{\omega_{t_1}\omega_{t_2}}$$

and

$$Q = \frac{\sqrt{F_{11}F_{22} - F_{12}F_{21}}}{(F_{11}\sqrt{\omega_{t_1}/\omega_{t_2}} + F_{22}\sqrt{\omega_{t_2}/\omega_{t_1}})} \tag{9.12}$$

and F_{11}, F_{12}, F_{21}, F_{22}, f_{1i}, f_{2i} are functions of resistor ratios which, when suitably chosen, allow realization of a bandpass or lowpass filter. At least three resistors are necessary and realization of an allpass or notch filter requires[20] an additional resistor-divider network.

It is easy to be misled by the apparent attractions of active-R networks. In practice, they are restricted to higher-frequency operation unless very large resistor ratios are employed. Most serious is their pronounced dependence on gain-bandwidth product which must be accurately measured before design and closely controlled thereafter (see Sec. 9.3). Indeed, the whole philosophy merely transfers the specification problem from an external capacitor to a less accessible internal capacitor. Single-amplifier realizations are only possible by introducing[21] an external capacitor as a partial active-R structure.

9.2.3 Dynamic range

As in the conventional communications sense, *dynamic range* expresses the lower and upper limits of operable signal level. The former is determined[22] by the allowable noise level in the filter while the latter is set by the incidence of distortion at large-signal excitations. The nature of this upper limit also depends on the frequency as is characterized by the two non-linear limiters in the block diagram of Fig. 9.7. That at the output is dominant at low frequencies where the linear block of Eq. (9.3) still provides high gains and is bounded by the maximum output-voltage swing V_{pp} of about 30 V peak to peak for common operational amplifiers. As frequency increases, the bound on the output voltage V_{out} is approximately expressed[23] by

$$V_{out} \leqslant \frac{V_{pp}}{\sqrt{1 + (f/f_T)^2}} \qquad (9.13)$$

where f_T is the *large-signal bandwidth*, i.e., the frequency band over which maximum output voltage swing is available.

At high frequencies, where the linear-block gain is significantly reduced, limiting transfers to the input saturation-type nonlinearity. This is measured[24] by the *slew rate* ρ, i.e., the maximum rate of change of the amplifier output

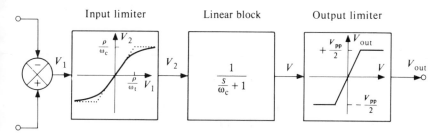

Fig. 9.7. Nonlinear model of operational amplifier

voltage with time and is a consequence of nonlinearities inherent in the junctions of the constituent transistors. It is manifested by triangularization of signal waveforms at higher frequencies because the input sinusoid is limited and then integrated by the linear block of the model. Long before such distortion occurs, there are anomalies in the frequency-response behaviour of high-Q active filters. These can be investigated numerically in terms of describing functions[25,26] when a linear approximation to the non-linear (dotted) curve is made as in Fig. 9.7. The output saturation level is given by

$$V_2 = \frac{\rho}{\omega_c} \qquad (9.14)$$

In extreme cases, the phenomenon may lead to jump resonance[25] characterized by the overhang response in Fig. 9.5 augmenting the linear imperfection effects. Many techniques are available for avoidance of jump resonance and limit cycles, but it is usually more practical to develop the limits on the performance of the filter (see Sec. 9.2.5) and redesign if necessary.

9.2.4 Passive-component parasitics

In practice, passive components are also non-ideal, as indicated by the lossy inductor model of Fig. 4.2. Similarly, capacitors have dielectric losses which may be simulated[27] by resistance while resistors have distributed capacitance[28] as described in Sec. 3.5. Usually, such parasitic effects only dominate filter performance at frequencies above 1 MHz where stray components also become significant.

However, for high-Q realizations, even at lower frequencies, the influence of capacitor losses can be quite marked.[27] Adopting the dissipative capacitor model depicted in Fig. 9.8, where $R_p \gg R_s$, then the terminal admittance is

$$Y = \left(\omega_3 \cdot \frac{s + \omega_1}{s + \omega_3} \right) C \qquad (9.15)$$

with frequencies as defined in Fig. 9.9. Replacing operator s by

$$\omega_3 \cdot \frac{(s + \omega_1)}{(s + \omega_3)}$$

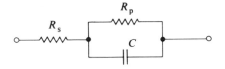

Fig. 9.8. Circuit model for dissipative capacitor

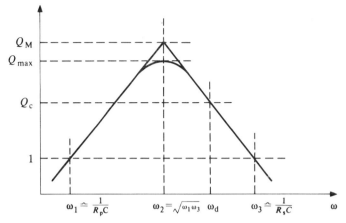

Fig. 9.9. Q plot of dissipative capacitor

in the denominator of second-order function $D(s)$ of Eq. (4.6) and determining the actual value of selectivity achieved gives

$$Q_e \triangleq \frac{Q}{1 + 2Q/Q_c} \qquad (9.16)$$

where

$$Q_c \triangleq \frac{Q_M}{\omega_2/\omega_d + \omega_d/\omega_2}$$

The importance of capacitor loss is now evident, since to ensure a deviation of less than 0.5 per cent from the lossless design Q requires that Q_c (see Table 9.2) exceed Q by at least two orders. The shift in centre frequency ω_0 is much less pronounced. This treatment assumes identical capacitances and neglects the frequency dependence of the model elements. In practice, whether the dielectric is non-polar (e.g., glass, mica), plastic (e.g. polystyrene), or integrated (e.g. tantalum oxide), the parameters are functions of frequency as yet not characterized and applied to filters.

9.2.5 Design limits

The device imperfections discussed above essentially depend on the available technology and determine the range of operation of a given filter. In order to illustrate simultaneously the limitations imposed on filter performance by each practical factor, the Q-frequency plane[29] is now adopted. As pointed out in Sec. 9.2.1, the linear imperfections amount to a limit on stability which is obtained by requiring the first-order coefficient of the quadratic in Eq. (9.4) to be positive, yielding a Q–f_0 contour for particular values of imperfections. This is

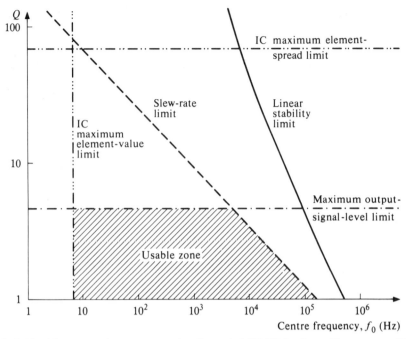

Fig. 9.10. *Q*-frequency plane representation for typical MLFB bandpass filter at a specified signal level

indicated by the full line in Fig. 9.10 for a typical multiple-loop negative-feedback (MLFB) bandpass filter design (see Table 7.3) with $\omega_t = 2 \times 10^6$ rad s^{-1}, $R_{in} = 1$ MΩ and $R_{out} = 100$ Ω. In defining the non-linear bounds, it is necessary to specify the input signal voltage of Eq. (9.13) at low frequencies. The condition for pure sine-wave signals at higher frequencies is expressed[30] by the peak output voltage

$$\hat{V}_{out} = K_0 \hat{V}_{in} < \frac{\rho}{\omega_0} \qquad (9.17)$$

If centreband gain K_0 is a function of Q, the dashed contour for avoidance of slew-rate limiting is then readily deduced for a given peak signal level $\hat{V}_{in}$ of 1 V. In this example, it is seen to set a more severe limit than does linear stability, but remember that it can be considerably relaxed for lower input voltages and by simple redesign of the same circuit configuration so that K_0 is independent of Q. This is achieved by requiring a wider component spread, introducing further technological limits. Imposing extreme integrated-circuit ratios for present practical thin-film resistors from Table 9.2 gives the upper horizontal limit in Fig. 9.10 while the maximum capacitance value dictates the lower vertical limit. The plane now delineates a usable zone which constitutes a basis of comparison between different designs and different types of active filter.

PRACTICAL CONSIDERATIONS

The question next arises as to how the performance limits of a filter using a non-ideal operational amplifier can be extended. There are two approaches:

(a) *External compensation* applied by additional exterior components specifically to desensitize the filter, i.e., active-passive counterbalancing by arranging[31] the transfer function of a simple passive RC section to effectively cancel $A(s)$. Alternatively, the effects of amplifier gain-bandwidth product can be cancelled by using closely matched amplifiers as in the GIC-derived circuit of Table 8.5. Such techniques naturally interact with the inherent saturation non-linearities so as to further limit the allowable voltage signal level. This may necessitate judicious introduction[32] of non-linear compensating devices (e.g., diodes) in the feedback path.

(b) *Improved amplifier design* with particular regard to increasing the bandwidth. However, this is usually offset[33] by deterioration in other properties such as power consumption. It is also necessary to minimize the size of, and therefore the area occupied by, the internal compensating capacitor by techniques such as transconductance cancellation. Transconductance can be reduced by introducing emitter-degeneration resistors in the input stage of the operational amplifier thereby enhancing slew rate at the detriment of offset voltage. High input impedance can now be achieved by FET input stages requiring BIMOS (bipolar and MOS on the same substrate) technology as in the TL084 or CA3140 amplifiers.

Whichever of these approaches is adopted is fraught by trade-offs and compromises. The cost of extra components in (a) must be balanced against the cost of expensive amplifier design and production in (b). In an effort to evade these problems, resort can be made to predistortion (Appendix 2) or post-design trimming (Sec. 9.5).

9.3 Environmental susceptibility

The efficacy of the methods just discussed for extending the design limits of active filters depends upon close control of the circuit parameters in the face of ageing, drift in d.c. power supplies, and environmental changes of humidity and particularly temperature. Application of the third identity of Table 4.2 enables the sensitivity of the filter transfer function to variations in temperature θ (°C) to be simply related to the component sensitivities by

$$S_\theta^T = \sum_{i=1}^{n_R} S_{R_i}^T S_\theta^{R_i} + \sum_{i=1}^{n_C} S_{C_i}^T S_\theta^{C_i} + \sum_{i=1}^{n_K} S_{K_i}^T S_\theta^{K_i} \qquad (9.18)$$

When all components of the same type are using identical processes and materials, as is usual in integrated circuits, then

$$S_\theta^{C_i} \simeq S_\theta^{C_{i+1}} \simeq S_\theta^C, \quad \text{etc.}$$

may be regarded as average component sensitivity[34] and factorized from Eq. (9.18). At the same time, Eq. (4.67) for capacitive reactance only indicates that

$$\sum_{i=1}^{n_C} S_{C_i}^T = S_s^T \tag{9.19}$$

while Eq. (4.65) for a transfer function ($m = 0$) also gives

$$\sum_{i=1}^{n} S_{R_i}^T = S_s^T \tag{9.20}$$

On substituting these expressions, Eq. (9.18) becomes

$$S_\theta^T = S_s^T (S_\theta^R + S_\theta^C) + S_\theta^K \sum_{i=1}^{n_K} S_{K_i}^T \tag{9.21}$$

demonstrating that the temperature sensitivity with regard to passive components is dependent primarily on the transfer function and materials used. Only the active-component contribution is dependent on the realization configuration. Of course, this result is not valid if widely differing materials are employed in different portions of the circuit.

From this finding has emerged the common design philosophy[35] of desensitizing by matching the temperature coefficient of resistance (TCR) equal and opposite to the temperature coefficient of capacitance (TCC) while arranging that the dimensionless gain parameters K_i are independent of temperature. Inspection of Tables 9.2 and 9.3 shows that this is feasible, especially for thin-film fabrication with highly correlated deviations in component values. Hence higher values of temperature coefficient are not necessarily undesirable. Furthermore, if the dimensionless parameters are merely controlled gains defined by mutually tracking resistors, then they are effectively insensitive to temperature. However, in circumstances when the amplifier gain-bandwidth product affects performance, its temperature dependence is not so easily zeroed. One method[36] is for the manufacturer to adjust the amplifier internal compensation by using the negative temperature coefficient of the stabilizing capacitor C_c to offset that of the open-loop gain. Alternatively, acknowledging that the objective is to reduce the overall thermal drift of the filter, the temperature variations of the active-device parameters can be designed[37] to cancel those of the external resistors and capacitors. In externally compensated amplifiers described in Sec. 9.2.5, neglecting R_{in} and

PRACTICAL CONSIDERATIONS

R_{out}, the transfer function can be expressed as

$$T(s) = \frac{N(s)}{s/\omega_t - sR_eC_e + D'(s)} \qquad (9.22)$$

where R_e and C_e are the external compensating elements. The relative variation (see Eq. (4.56)) is

$$\frac{\Delta T}{T} = \frac{\Delta R_e}{R_e} \cdot S^T_{R_e} + \frac{\Delta C_e}{C_e} \cdot S^T_{C_e} + \frac{\Delta \omega_t}{\omega_t} \cdot S^T_{\omega_t} \qquad (9.23)$$

which, when zeroed, gives the compensation condition

$$\frac{\Delta \omega_t}{\omega_t} = -\left(\frac{\Delta R_e}{R_e} + \frac{\Delta C_e}{C_e}\right) \qquad (9.24)$$

This implies that the external compensating components have summed temperature coefficients equal and opposite to that of the gain-bandwidth product. Values of the latter for the μA741 amplifier are quoted[38] in the range −2000 to −3000 p.p.m. °C⁻¹. Although discrete and monolithic passive components are capable of meeting this requirement, present film technology[6,7] is strained. But it must be remembered that thin-film research has been hitherto directed towards achieving either absolutely low[6] temperature coefficients or matched[5] TCC and TCR rather than to the specific attainment of higher controlled coefficients.

9.4 Performance comparisons

The policy adopted in this text of presenting filter configurations capable of satisfactory performance in other than purely laboratory situations reflects the fact that several structures discussed in earlier chapters are manufactured commercially. The important decision task of selecting the circuit most suited to particular operational requirements is facilitated by the comparative tables and perspective charts at the ends of Chapters 6, 7, and 8. Optimal choice necessitates further considerations such as complexity, substrate space or area restrictions, reliability, d.c. supply requirements, power dissipation, and operating frequency range. These are in addition to the factors discussed in Secs. 9.1 to 9.3 and have a direct influence on cost, which ultimately decides the feasibility of mass production. Consequently, no definitive statements can be made as to which filter type is best in all circumstances, only guidelines to choice being suggested.

9.4.1 Sensitivity considerations

With respect to sensitivity, the overriding concern is to choose a filter circuit whose amplitude/frequency response is affected as little as possible by parameter variations. In this regard, Eq. (4.39) has indicated that the transfer function is $2Q$ times more sensitive to changes in ω_0 than to changes in Q at the half-power frequencies of a second-order section. It is sensible, then, to select[39] a configuration with negligible frequency sensitivity to the worst-tolerance element (often the active device). This is true for almost all circuits previously tabulated and is attended by passive frequency sensitivities $\leqslant 0.5$. Turning to the Q sensitivities, it is convenient to distinguish between the various features affecting circuit performance, only some of which relate to the structure itself. This is apparent for hybrid-integrated filters whose performances can be assessed by developing the variation of network function T with respect to A by Eq. (4.56) as

$$V_A^T = (A S_A^T) \frac{\Delta A}{A^2} \tag{9.25}$$

where the figure of merit $A S_A^T$ is termed[40] the *gain-sensitivity product* (GSP) depending only on the network structure. On the other hand, the factor $\Delta A / A^2$ clearly refers only to the particular amplifier employed. Since the centre frequency is usually determined by tracked resistor and capacitor products, the associated $A S_A^{\omega_0}$ depends upon the technology used to implement the filter and network interest is justifiably concentrated on $A S_A^Q$. When the latter is applied to each of the entries in Tables 6.7, 7.3, and those relevant in 8.5, all the negative-gain configurations have GSP $\propto Q^2$ (and are thereby limited to relatively low-selectivity applications) except for those incorporating twin-T sections which, along with most other configurations, have GSP $\propto Q$. Considered in conjunction with the corresponding passive Q sensitivities, this advantage is counterbalanced, illustrating the fundamental impossibility[16] of simultaneously reducing both active and passive sensitivities. The product $|A S_A^Q| |S_{RC}^Q|_{max} \propto Q^2$ in each case, except for the unity-gain mixed-source circuit of Fig. 6.16 with its conspicuously low GSP independent of Q in addition to low passive sensitivities.

It is interesting to compare the GSP results with those obtained from a development of the multiparameter statistical sensitivity (MPS) definition in Sec. 4.3.3. For the second-order transfer function of Eq. (4.6) with high Q, considered in the $\Omega_1 = 1 - 1/2Q$ to $\Omega_2 = 1 + 1/2Q$ bandpass region and assuming $N(s)$ has no zeros in the vicinity of the denominator roots, then the sensitivity measure of Eq. (4.75) may be expressed[41] in terms of the denominator function as

$$M(x) = \int_{\Omega=1}^{|\Omega_2|} \sum_{i=1}^{n} |S_{x_i}^D|^2 \sigma_{x_i}^2 \, d\Omega \tag{9.26}$$

This is valid under a linear approximation with zero component cross correlations.

Interpreting in terms of the frequency and Q sensitivities of Eqs. (4.33) and (4.34) gives a convenient means of obtaining a lower bound on $M(x)$. In the case of parameter $x_i = A$, this produces the criterion

$$|S_A^p| = \sqrt{(S_A^Q)^2 + 4Q^2(S_A^{\omega_0})^2} \tag{9.27}$$

which is evaluated at $\Omega = \Omega_0 = 1$ and indicates the same overall dependence on Q as the GSP for all cases except the single- and mixed-source negative controlled-gain circuits. The latter, with their finite $S_A^{\omega_0}$, invalidate the initial premise on which GSP is based. Nevertheless, for all circuits with zero frequency sensitivity to amplifier gain, GSP may be invoked as a convenient substitute for the criterion developed from the all-embracing MPS.

9.4.2 Frequency limits

Good filtering performance in the upper audio band is of interest in applications such as 'hi-fi' equipment and ultrasonics while still higher-frequency operation is required if active filters are to entirely replace LC networks in order to form a technologically compatible system. The order of the upper frequency limit of operation for each filter circuit may be assessed by substituting the appropriate design conditions presented in Tables 6.5, 7.3, and 8.5 into the corresponding ω_0 expression to reveal the Q dependencies. Although the centre-frequency expressions for the positive-gain single-source and the multiple-loop-feedback configurations are superficially alike, there are different component-choice implications, resulting in the former being independent of Q while the latter is inversely proportional to Q. This appears to confirm the well-known[35] high-frequency limitation of selective MLFB single-amplifier filters. However, it should be remembered that this comparison is based on only one set of design conditions. Had the element choice $G_1 = G_2/(3Q - 1) \rightarrow G_3/3Q$, $C_1 = C_2$, and $K = (9Q - 3)/(3Q - 1) \simeq 3$ been made for Fig. 6.2 and $G_1 = G_2/(4Q^2 - 1) = G_5/2$, $C_3 = C_4$ for Fig. 7.5, then, in each case, $\omega_0 \propto Q$. This implies that both circuit configurations are capable of much improved high-frequency behaviour by redesign, but at the expense of an inconvenient component spread, i.e., constituting a fundamental trade-off. Similar observations can be made regarding most of the other tabulated filter structures.

As discussed in Sec. 9.2, high-frequency non-linear limits are set by the slew rate, which is most restricting[23] when centreband gain K_0 is a function of Q, a fact related to the centre-frequency dependence just discussed. Slew-rate limiting is also ameliorated by using higher-power-rating amplifiers, i.e., yet another design trade-off, as witnessed in Table 9.1. But, at small signal levels, the dominant high-frequency restriction remains the effect of the finite gain-bandwidth product of the amplifiers used. The relative importance of GBWP on the sensitivities of some popular active filters may be judged by Table 9.4, which again displays[16] the superiority of positive-gain over negative-gain (MLFB) single-amplifier configurations. Among the multi-amplifier circuits, the state-variable two-integrator loop has a pronounced $S^Q_{\omega_t}$, severely limiting its application for selective high-frequency specifications. The allpass pair[17] of Fig. 9.4, formed by connecting the first-order stages of Fig. 7.10(a) in a loop, is much better in this respect since the sensitivity problem is effectively transferred to the more readily controlled gain-defining resistors. The apparently outstanding δQ characteristic of the GIC-derived circuit is dependent on close matching of the amplifiers to balance out otherwise large perturbations. Moreover, the advantage of this active compensation is not so great since second-order effects soon dominate and Q-enhancement techniques may be preferable. Direct use of the ω_t pole itself (if it can be controlled, see Sec. 9.2.2) allows natural extension of operating frequency. Other issues to consider are that controlled-source realizations (particularly unity-gain) often allow direct use of transistors with their wider frequency response than OA's while distributed-lumped networks[42] (DLAN's) overcome many passive parasitic problems at high frequencies.

At the lower end of the audio band, where the OA may more justifiably be assumed ideal and active realizations are obviously superior to their passive counterparts, design is facilitated by resort to the simplified relations of earlier chapters. As discussed in Sec. 9.2.5, the operating limit at very low frequencies depends on the technology adopted. If R_m and C_m are the total substrate resistance and capacitance, respectively, then the normalized lower frequency limit $\omega_0 R_m C_m$ can be adopted as a figure of merit for comparing different filter realizations. Although it is dependent on design conditions, some conclusions can be drawn for high Q. For example, the positive-gain structures without controlled feedback have $\omega_0 R_m C_m$ independent of Q, as do the modified MLFB and generalized DI configurations. However, the majority of circuits, including the state-variable and GIC-derived circuits in Table 8.5 and all twin-T configurations, indicate $\omega_0 R_m C_m \propto Q$. The lower limit becomes restricted to order Q^2 and Q^3 for the mixed-source unity-gain and DVCCS/DVCVS, respectively. These bounds should be taken into account with those at upper frequencies discussed in the previous paragraph. Approximate numerical limits for a sample of filters evaluated in the case of second-order bandpass realization at $Q = 10$, subject to the integrated-circuit restrictions $R_m = 2.5$ MΩ and $C_m = 50$ nF, are

multiple-loop feedback,	53 Hz
state-variable,	53 Hz

PRACTICAL CONSIDERATIONS

Hence, under the same hybrid constraints, the multi-amplifier configurations can be more restricted in this regard than canonical realizations. For more sophisticated responses, including the notch, extension of low-frequency limits requires consideration of the ladder arrangements of Chapter 8, for which excellent simulated-inductance characteristics are possible. The attainment of still lower frequencies for certain instrumentation applications necessitates connection of external lumped capacitors, 10 µF for example allowing limits in the above cases of 0.05 Hz and 0.265 Hz. This represents a physical limit because of the difficulty of obtaining larger values of precision components without either excessive physical size or prohibitive cost. Commercial laboratory filters (see Sec. 9.7), being designed with discrete components, do not suffer from substrate constraints and, using metal-film resistors and metallized polycarbonate capacitors, are capable of operation at 0.001 Hz.

9.4.3 Space occupied

Small physical size is obviously of prime importance in microsystem filters, besides being directly related to manufacturing and materials costs. An index of the space occupied (or total substrate area in the case of integrated circuits) is obtained[43] from the product of summed resistor and capacitor values, namely $(\sum C_i)(\sum R_i)$. For a given operating frequency, this is dependent on two factors:

(a) *Component count*, which is an important consideration in both hybrid and discrete forms of construction. In the one case, economy in components reduces the substrate area and the need for elaborate interconnection patterns while, in the other, it minimizes both the expense of precision discrete elements and the number of soldered joints. In either case, reliability is enhanced. Failing the development of adequate monolithic passive components, thin-film distributed sections offer a favourable solution in this regard. Of course, distributed resistance and capacitance (see Sec. 3.6) are difficult to fabricate accurately, but the manner in which they influence expressions of interest means[34] that their tolerances are less critical than other adjustable lumped parameters. Perhaps a greater inconvenience is that $\overline{RC}$ transmission lines do not exhibit rational second-order transfer functions for compatibility in the high-order systems of Sec. 9.7. Biquadratics with non-finite transmission zeros can be canonically realized, for example, omitting resistor R_2 in the multiple-loop-feedback circuit of Table 7.1. However, this is achieved at the expense of inflexible design and loss of tuning facility and, furthermore, the extra resistor is small in value. Controlled-source realizations, except in the unity-gain varieties, necessitate additional gain-defining resistors. This near-canonical feature of MLFB circuits helps to counterbalance its otherwise deleterious properties and it is still a popular choice for non-exacting filter specifications, particularly in instrumentation when the non-inverting OA input can be utilized for variable d.c. bias. By contrast, mixed-source, differential-input, and all twin-T realizations are extravagant in components and may be cost-prohibitive in discrete assemblies, whereas in hybrid form, superior temperature tracking and amenability to laser trimming render them feasible. When finite transmission zeros are required, there is little to choose in component count between the various second-order designs considered.

(b) *Component spread*, which is of more concern in hybrid-integrated realizations on account of the limited ranges of resistance and capacitance values indicated in Tables 9.2 and 9.3 together with the inferior tracking tolerances of components with widely differing values. In fact, for high selectivities, the maximum component spread m has in all but one case the same Q dependence as $(\sum C_i)(\sum R_i)$. This appears to be true for all design conditions provided that the smallest resistor is included in the basic time constant and implies that the latter index adds little to the information contained in the spread itself. Substrate area requirements can therefore be entirely assessed by means of maximum component spread, which can be found listed in comparison Tables 6.5, 7.3, and 8.5 for the specified design conditions. In comparing the circuits of Table 8.5, it should be remembered that the nominal spread of unity in several cases excludes the component range inside the necessary converters, which may be of order Q. The related bridged-T and simple multiple-loop feedback circuits of Table 7.3 appear to be precluded for integrated implementation except at low Q's. A passive spread of Q proportions can usually be entertained even at high selectivities, especially if it refers to a single element in relation to a group of comparable-valued components, as in the state-variable design. However, it should be weighed alongside other properties and a possible overall figure of merit based on Eq. (9.27) is

$$\Lambda = \sqrt{(2QS_A^{\omega_0})^2 + (mS_{RC}^Q S_A^Q)^2} \qquad (9.28)$$

Even this neglects many considerations; the apparent superlative performance of the mixed-source unity-gain filter of Fig. 6.16 disregarding, for example, the technological inconvenience of employing two types of source and the complexity in terms of passive and active component count, to say nothing of likely power dissipation and noise-induced instability. For simplicity, the positive-gain Sallen and Key circuit has much to recommend it as a second-order section. But, in general, more flexible high-frequency operation can be obtained with the allpass pair of Fig. 9.4 (with close control of resistors) or the GIC-derived structure of Fig. 8.23 (with close matching of OA's).

9.5 Tuning

The initial production tolerances discussed in Sec. 9.1 mean that post-design adjustment is an essential process step in meeting close specifications, especially for high-density hybrid integrated-circuit filters. While predistortion (see Appendix 2) furnishes an alternative technique for compensating the effects of component imperfections, only subsequent tuning can take account of production spreads. The particular tuning method employed depends upon the function to be realized, the network structure used and the technology of implementation.

9.5.1 Functional tuning

This is performed with the circuit in its operational mode, entailing[44] parameter modification by change of prescribed components at frequencies of known phase shift. The filter properties to be tuned are best observed using phase characteristics as a barometer since these are subject to greater changes than the amplitude in the region of ω_0.

Ideally, each parameter of interest is controlled in a non-interactive fashion by changes in a single circuit element which, in the case of hybrid construction, must be a resistor. Thus, for a network having design parameters δ_1, δ_2, and δ_3 set by variation of circuit elements x_1, x_2, and x_3, development of Eq. (4.17) gives

$$\begin{bmatrix} \Delta\delta_1 \\ \Delta\delta_2 \\ \Delta\delta_3 \end{bmatrix} = \begin{bmatrix} S_{11} & 0 & 0 \\ 0 & S_{22} & 0 \\ 0 & 0 & S_{33} \end{bmatrix} \begin{bmatrix} \Delta x_1 \\ \Delta x_2 \\ \Delta x_3 \end{bmatrix} \qquad (9.29)$$

indicating the need for a diagonal sensitivity matrix. Unfortunately, this is rarely achieved in practice and often the best that can be hoped for is a triangular matrix

$$\begin{bmatrix} \Delta\delta_1 \\ \Delta\delta_2 \\ \Delta\delta_3 \end{bmatrix} = \begin{bmatrix} S_{11} & 0 & 0 \\ S_{21} & S_{22} & 0 \\ S_{31} & S_{32} & S_{33} \end{bmatrix} \begin{bmatrix} \Delta x_1 \\ \Delta x_2 \\ \Delta x_3 \end{bmatrix} \qquad (9.30)$$

suggesting a prescribed tuning sequence. In this case, adjustment of δ_1 by means of x_1 affects δ_2 and δ_3, but subsequent tuning of δ_2 by x_2 does not affect δ_1 nor, in turn, are either δ_1 or δ_2 affected when δ_3 is tuned by x_3. State-variable biquadratic filters fall into a similar category, providing a further reason for their popularity. For the circuit of Fig. 8.14, modified by the addition of R_{10} between earth and the junction of R_1 and R_2, the design parameters are

$$\omega_0 = \sqrt{\left(\frac{R}{R_3}\right)\frac{1}{C_1 C_2 R_8 R_9}}, \quad K_0 = \left|\frac{V_2}{V_1}\right|_{max} = \frac{R_2}{R_1}$$

and

$$Q = \frac{R_3(R_1 + R_2 + R_1 R_2/R_{10})}{R_1(R + R_3)} \sqrt{\frac{R}{R_3} \cdot \frac{C_1 R_8}{C_2 R_9}}$$

(9.31)

The hybrid tuning sequence corresponding to Eq. (9.30) is

x_1: R_8 and/or R_9 to set centre frequency, $\omega_0 = \delta_1$,
x_2: R_1 or R_2 to set maximum gain constant, $K_0 = \delta_2$,
x_3: R_{10} to set selectivity, $Q = \delta_3$.

As a general rule, analogue-simulation and other multiple-amplifier structures have a low degree of element interaction and are relatively easy to tune functionally. For single-amplifier arrangements, an iterative sequence is usually necessary, though this can be time consuming and, in some cases, non-convergent. For example, while circuits of the type in Fig. 6.22 have frequency and independent Q control by means of R_1 and R_4, respectively, adjustment of R_3 is necessary to preserve the null-adjustment. The latter, unfortunately, interacts with Q.

Transfer functions containing finite $j\omega$-axis transmission zeros, if realized directly by LC ladders or their active simulations, are admirably suited to functional tuning because of the non-interacting series or shunt resonant arms. However, if obtained by null-adjusted RC sections (whether the twin-T or analogous distributed notch of Fig. 3.14), they are necessarily more difficult to tune[45] than all-pole functions. Structures in which the zeros are formed by an output summing or difference amplifier can have this active element temporarily disconnected to allow separate adjustment of the poles. Furthermore, in many practical cases, the numerator and denominator expressions can, for the purposes of tuning, be regarded as non-interactive since their regions of significant phase change are sufficiently well separated.[9] This occurs if a_0 and b_0 in Eq. (6.16) differ by more than ten per cent.

The trial and error process of functional tuning for precision is inherently slow and, in the case of laser trimming (Sec. 9.1), finite settling times must be allowed between successive element adjustments. It is generally only recommended for simple systems and is further limited in hybrid IC iterations by the irreversibility of resistor trims.

9.5.2 Deterministic tuning

The time-consuming iterations mentioned above can be avoided by introducing[46,47] a predictive step comprising an initial circuit analysis. This has proved especially effective in large-scale hybrid production facilities accompanied by automated laser trimming,[48] in which circumstances it can be performed at many times the speed of functional tuning. In addition, it can be adopted for a much wider variety of structures.

In the most general case of deterministic tuning, exact design equations are formulated taking into account the frequency-dependent amplifier gains of Eq. (9.3) and capacitor losses contained in Eq. (9.15). With the circuit interconnection pattern as yet incomplete, preliminary measurements are made to establish exact values of the cited parameters. On substituting these into the design equations, a set of non-linear expressions for the circuit resistors emerges necessitating on-line computer solution. In the event of multiple roots, that closest to the design value is selected. Laser trimming is then applied to adjust the resistors within 0.1 per cent of prescription. However, after pinning the circuit connections and energizing *in situ*, the specification may not be met owing to unaccounted variations in the filter environment (e.g., switched input or output in a PCM channel-bank multiplexer). This necessitates further functional adjustment[47] by trimming a gain-defining resistor and thereby

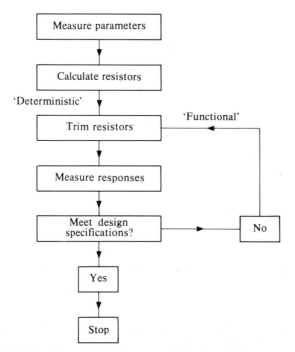

Fig. 9.11. Deterministic tuning procedure diagram with functional adjustment

compensating the nominal gain at a fixed frequency for manufacturing tolerances in switch 'on resistance' and holding capacitance. The whole sequence is outlined by the simplified block diagram of Fig. 9.11. A final trim at the 'yes' stage can be carried out to eradicate any previous measurement and adjustment errors as well as operational-amplifier variations.

For practical purposes, the capacitor shunt resistance R_p in Fig. 9.8 can often be ignored and, in the voice-frequency band, the amplifier regarded[47] as ideal. The design equations are then considerably simplified, resulting in a much faster tuning process together with reduction in software complexity. For example, the expressions for the single-source arrangement of Fig. 6.2 can then be simply reformed in terms of conductances as

$$G_1 = \frac{HC_1}{K}, \quad G_3 = \frac{C_1 C_2}{\omega_0^2 (G_1 + G_2)},$$

and

$$G_2^2(1 - K) + G_2 \left(2G_1 - KG_1 - \frac{\omega_0 C_1}{Q}\right)$$
$$+ \left[G_1^2 - \frac{\omega_0 G_1 C_1}{Q} + \omega_0^2 C_1 (C_1 + C_2)\right] = 0 \quad (9.32)$$

yielding specific values for G_1, G_2, and G_3 when the prescribed values for ω_0, Q, and H are inserted together with the measured values of C_1 and C_2. Nevertheless, the equations still require computational aid and, for more advanced transfer functions and highly interactive structures, the software can be formidable. The need for complex algorithms can be overcome[49] by estimating transfer-function coefficients after each resistor adjustment and hence recalculating trim resistors until a weighted sum of squares of the coefficient errors is minimized.

9.5.3 Automatic tuning

In applications where the input signal frequency varies over a wide range (as in vibrating-transducer instrumentation), the wide-bandwidth requirement is in conflict with effective noise rejection. This can be resolved by automatically adjusting the centre frequency of a selective bandpass filter so as to track the signal frequency. In order to preserve a constant bandwidth $B = \omega_0/Q$ rad s^{-1} as well as midband gain K_0, it is necessary to adopt an RC-active filter which allows frequency control by a single resistor not contained in the expressions for B and K_0. Inspection of Tables 6.5, 7.3, and 8.5 indicates this facility to be available in the state-variable, GIC-derived, and unity-gain mixed-source configurations. However, the only single-amplifier configuration with this

PRACTICAL CONSIDERATIONS

property is the bandpass multiple-loop feedback circuit since, for equal capacitors in Table 7.1,

$$\omega_0 = \sqrt{G_5(G_1 + G_2)}/C, \quad B = \frac{2G_5}{C}, \quad \text{and} \quad K_0 = \frac{-G_1}{2G_5} \qquad (9.33)$$

The latter two expressions are independent of G_2, which may therefore be used to tune ω_0. The automatic capability is added[50] by means of a phase-sensitive detector (see Fig. 9.12) which differentiates the input and output filter waveforms before comparison, gating, smoothing, and application to the FET acting as a voltage-variable resistor.

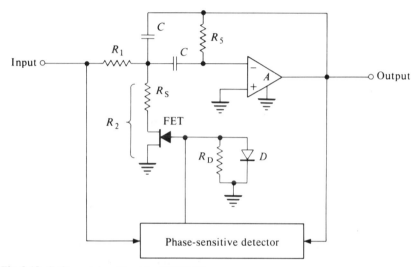

Fig. 9.12. Self-tuned signal-tracking MLFB filter

The system, in fact, comprises a phase-locked loop (PLL) which tracks a given signal while admitting noise only in a small bandwidth. PLL's have been advocated[51] as alternatives to RC-active filters generally for highly selective bandpass requirements with excellent noise immunity. Once 'lock' is acquired, tracking capability is close even in heavy noise, but the frequency-capturing problem can be more difficult. They are available in monolithic form needing minimal numbers of external components whose tolerances need not be tightly controlled. However, the technique is not readily amenable to realization of any prescribed response.

9.6 High-order filters

As explained in Chapter 2, real filtering problems usually require resort to high-order approximating functions. One approach to their realization has been hinted in Chapter 4, namely, by forming cascades of simpler sections. Another is represented by the techniques of Chapter 8, based mainly on passive ladder prototypes. It is the intention of this section to consider the practicalities of such methods as distinct to direct *RC*-active synthesis.

9.6.1 Cascaded systems

By its sheer simplicity, the concept of cascading networks exhibiting biquadratic transfer functions has much to commend it. Furthermore, second-order sections are readily realized using the methods of Chapters 6 and 7. In the event of an odd-order specification, the additional first-order factor may be realized by a single *RC* section or by associating it with one of the biquadratic blocks to form a third-order structure. An inherent feature is that variation in a given element affects only one pole-zero pair, a fact which has been demonstrated in Sec. 4.2.4 to imply reasonable sensitivity as well as easing the tuning process. The individual stages are readily isolated by relative impedance scaling but the siting of an amplifier at the output of each stage permits impedance levels to be chosen more flexibly.

Major influences on the overall performance of the entire filter are the *pairing of poles and zeros* to form the various biquadratic sections and the *positioning* of these sections in the cascade. In order to minimize distortion, best results are obtained when the poles and zeros are chosen so as to produce as flat an amplitude response as possible for each section in the frequency band of interest. Tabulated maximum–minimum function ratios for each pole-zero combination are prepared[52] and consulted for this purpose, generally resulting in their being chosen as close together as possible. From the point of view of sensitivity minimization, the effect of pole-zero pairing depends on the particular realization of constituent second-order sections. Under the assumption that minimizing the sensitivity of each biquadratic block does likewise for the whole system, then networks having only one forward path in their signal-flow graph (e.g., the single controlled-source circuit of Fig. 6.1 and those using the FEN system of Fig. 6.19) have sensitivities independent[53] of pole-zero pairing. On the other hand, networks having two forward paths (e.g., the state-variable realization of Fig. 8.14 and the generalized modification of the MLFB shown in Fig. 7.11) have sensitivities which *are* dependent on pole-zero pairing. In this case, it is found advisable to associate complex poles with the zeros most distant from them when decomposing the transfer function. This, of course, is in direct conflict with the usual requirement for distortion discussed above (though, as outlined in Sec. 9.5.1, it can also simplify tuning), but no general rule can be stated without reference to a specific realization.

Broad guidelines are available for the positioning of individual types of second-order section in the network cascade. Both the first and last stages of a high-order filter should preferably be of the lowpass or bandpass form; that at the input to avoid the slew-rate problems of Sec. 9.2.3 by keeping higher frequencies away from amplifiers and that at the output to suppress amplifier noise. Also, wherever possible, a stage with a peaked frequency response should be preceded by stages providing attenuation in the region of the centre frequency in order to obtain good large-signal-handling capabilities.

9.6.2 Multiple-loop systems

The signal-flow graph for the general state-variable system in Fig. 8.12(a) can be further developed to include transfer functions $T_j(s)$ in place of simple integrators. The resulting high-order filter system illustrated in Fig. 9.13

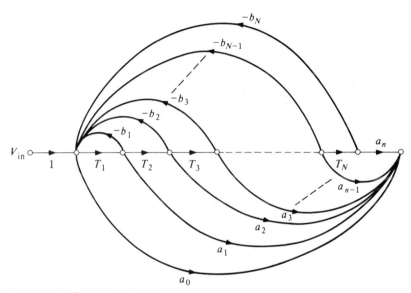

Fig. 9.13. Signal-flow graph for FLF system

employs *follow-the-leader feedback* (FLF), so named[54] because each successive feedback path b_k follows the lead of that preceding in 'peeling off'. Finite transmission zeros are generated by the summation of feed-forward paths a_i and the voltage transfer ratio of the system is readily derived as

$$T = \frac{V_{out}}{V_{in}} = \frac{a_0 + \sum_{i=1}^{N} a_i \prod_{j=1}^{i} T_j}{1 + \sum_{k=1}^{N} b_k \prod_{j=1}^{k} T_j} \qquad (9.34)$$

Substitution, expansion, and coefficient matching are used to effect a design solution. Though first-order stages offer more flexibility, it is convenient to utilize the ready-synthesized active second-order sections. However, this restricts design to geometrically symmetrical bandpass and bandstop filters unless resort is made to *generalized follow-the-leader* (GFLF) techniques[55] which allow any transfer function by incorporating general biquadratic sections or *biquads*.

Various useful cases arise by simplifying Fig. 9.13. For example, removal of the first feedback path b_1 results[56] in the *shifted-companion form* (SCF) in which all but the first biquad have identical pole Q's and design is aided by recursive formulae. The system may be further reduced by the removal[57] of all feedforward paths to form the *primary resonator block* (PRB) arrangement. Though clearly capable of realizing only all-pole functions, it has the practical advantage of utilizing identical biquads. Yet another variation is to remove the feedback paths but retain the feedforward paths, thereby forming the *inverse*[58] *follow-the-leader* (IFLF) configuration, which may be more difficult to tune but has superior dynamic-range properties compared with FLF. Dynamic range can be maximized[59] by exploiting unconstrained gain constants to equalize amplifier output voltages and ensure that signal levels throughout the filter are as large as possible.

In comparison with cascade systems, multiple-loop FLF techniques can reduce the highest pole Q required in the implementation of a given transfer function and consequently may demand only single rather than multiple-amplifier section realizations. This may, however, be achieved at the expense of an extra summing amplifier unless the feedback paths can be fed into the first section. Also, FLF offers many degrees of freedom and optimal solutions are not well defined. But the pre-eminent advantage lies in sensitivity performance improvement, especially when highly selective bandpass designs are considered. This is to be expected since systems with overall loop feedback are demonstrably[60] less sensitive than those with cascaded inner loops. It is the other philosophy of multiple looping, namely leapfrog (Sec. 8.2.1), which appears[61] to offer the best overall sensitivity. However, this usually requires modification as MLF and, the associated optimization procedure being non-trivial, FLF offers a practical alternative with systematic optimization and tuning together with modest fabrication costs in PRB form.

Further remarks of interest are that the cascade is, in fact, a special case of FLF and compromise solutions retaining the modularity of the former are possible by using multiple-loop sections to realize finite transmission zeros[62] and high-Q terms[55] as part of a cascade. Finally, it should be observed that the multiple-loop flow diagrams are identical to those employed in the direct form[63] of canonical digital-filter implementation. However, the requirement for precise control of feedforward and feedback usually makes other digital structures preferable.

9.7 Commercial availability

Active filters are now manufactured by a substantial number of companies, as evidenced by the sample presented in Table 9.5. These are usually supplied in

PRACTICAL CONSIDERATIONS

Table 9.5 Some manufacturers of active filters

Company	Type of active filter					Type of technology
	Fixed frequency	Tunable		Multi-channel and test	Special purpose	
		State variable	Other tuned			
AP Circuit Corp. (USA)	●●●					D, T₁, T₂
Barr & Stroud (UK)	●●●					D
Beckman Instrument (USA)		●●●●	●	●●		T₁, T₂
BEI (USA)^f	●●●●					T₁
Burr-Brown (USA)^a		●			●●●	D, T₁, T₂
Data Delay Devices (USA)					●●●	T₁
De Coursey (USA)					●●	D
Frequency Devices (USA)^b			●		●	D
General Instrument (USA)^d				●●●		T₁
Ithaco (USA)					●	D
Kemo (UK)					●●	D
Krohn-Hite (USA)		●●			●●	D
MITEL (Canada)						T₁
National Semiconductor (USA)						D
Optical Electronics (USA)					●	D
Rockland Systems (USA)^e				●		D
Sprague Electric (USA)						T₁
Welwyn Electric (UK)^c	●					T₁

Note: D = discrete, T₁ = thick film, T₂ = thin film.

hermetically sealed packages, perhaps housing a number of substrates. The range of products is conveniently categorized as follows:

1. *Fixed-frequency filters* as discussed in earlier chapters. These are available for all functional types, popular approximating characteristics being Bessel, Butterworth, and Chebyshev. The module is customarily specified as two-, four-, six-, or eight-pole with cut-off frequency typically between 1 Hz and 20 kHz, though as low as 0.01 Hz for some packages. Network structures are usually based on the positive-gain controlled-source type (Sec. 6.1.1) for the reasons stated in Sec. 9.4. Adjustment of centre frequency and Q factor is effected by the manufacturer, e.g., resistor-padding tuning[a]† as explained in Sec. 9.5. In some cases,[b] single-resistor control of selectivity is possible or, for bandstop filters, a notch-trim resistor is provided. Other commercial structures in this category are the lowpass version of the Lim circuit[c] discussed in Sec. 6.1.2 and the multiple-loop feedback arrangement[d] of Sec. 7.2.

2. *Tunable filters* in which the response is externally adjusted by resistive, voltage, or digital tuning. The most widely available type is that based on the state-variable two-integrator loop of Fig. 8.14, capable of simultaneously supplying lowpass, highpass, and bandpass outputs. The circuit is produced with the key frequency and Q-determining resistors (R_8, R_9, and R_2) omitted and is commonly known as the *universal active filter* (UAF). Most manufactured varieties contain an uncommitted fourth amplifier, permitting realization of notch and biquadratic functions as well as providing additional gain. Other forms of tunable filter based on the UAF include:[b]

    (a) BCD/binary switch-programmable filters displaying fourth-order lowpass Bessel and Butterworth characteristics whose cut-off frequency is programmed by externally interconnecting (mechanically or electrically) appropriate terminals of the filter module;

    (b) voltage-tunable notch filters[64] incorporating analogue multipliers or sampling switches to effect variation of the system time constants.

3. *Multi-channel and test filters* for use in signal processing, commissioning, and specifying fixed-frequency filters. Again, they are usually based on the positive-gain controlled-source structure exploiting a particular practical advantage arising from the fact that the same element values (with capacitors and resistors interchanged by a panel switch) can be used to render both the highpass and lowpass Butterworth characteristic. Bandpass responses may be obtained by a cascade connection of highpass and

† Text references to manufacturers are denoted by a single superscript italic letter which relates to the coded entries of Table 9.5.

lowpass channels, bandstop by parallel connection. A more sophisticated arrangement is available[e] with capacity for sixteen independent channels and provision for local, preset or remote programming with read-out facilities.

Tuning of test filters is frequently continuous or multiposition digital, though some models provide only selected centre or cut-off frequencies over the operating range. Capacitors are fixed for each range and approximate resistor values are selected by ganged switches or potentiometers, according to the desired cut-off frequency. The tuning range may be as narrow as a single decade, is typically five decades and perhaps as wide as eight decades. This covers the spectrum from 0.001 Hz to 3 MHz though, with currently available OA's, gain-bandwidth products naturally result in poorer accuracy at the high-frequency limit. Because of the wide operating range, resistor values are required with a spread of $10^4:1$ and special care must be taken to ensure low OA bias currents, necessitating adoption of bi-FET devices.

4. *Special-purpose filters* mainly for applications in telephone and other communication channels whose enormous potential market has influenced specifications. In the main, they are manufactured in-house or are custom-designed. Of those available commercially, tone-receiver filter systems are the most common. An array of eight bandpass filters having centre frequencies in the range 697 Hz to 1633 Hz is representative of this category. The individual Q factors are preset at 22 (± 10 per cent) with a centre-frequency tolerance of ± 0.5 per cent. The UAF is well suited to this application since it can provide stable Q factors with low sensitivity to passive and active elements. For less exacting applications, a single-amplifier realization may be adequate (see Secs. 6.1 and 7.3.3). A fifth-order Cauer (elliptic) circuit satisfies the specification for the D3 channel-bank filter and is produced by a number of companies. High-quality lowpass performance[47] is obtained from a cascade of the grounded twin-T circuits of Fig. 6.8, as illustrated in Plate 1. Alternatively, there is an FDNR realization[f] of the *LCR* lowpass prototype (Sec. 8.3.2). In addition, active filters may be operated as oscillators when designed to satisfy the $Q_m = \infty$ condition in Sec. 4.1.1.

9.8 Future developments

Several avenues for future research have been revealed in the exposition of this text. Among the more empyreal may be listed the need for a two-port energy theory capable of reconciling the realizability, selectivity, and sensitivity of both passive and active transmission networks. On a more practical plane, the apparent promise of some structures considered in earlier chapters such as the mixed-source unity-gain configuration of Fig. 6.16 requires further investigation. Manipulation of pathological-element models could yield new practical circuits even though this area appears so exhaustively explored. High-order filters such as the follow-the-leader system of Fig. 9.13 require more sensitivity-optimization studies. In this regard may be stated the desirability of improving the multiparameter statistical measure of Sec. 4.3.3 to relate more closely to production yield. There is a need for greater awareness of the effects of component imperfections on filter responses with development of effective compensation techniques to extend the range of filtering. In the case of device nonlinearities, increased understanding of the underlying phenomena may eventually permit their use in forming a specified response.

Progress in integrated technology is now allowing[65] lower-cost thick-film resistor connections of base metals while further improvements in the manufacture of thick-film capacitors would considerably accelerate hybrid applications. The most significant advance in refining microelectronic device specifications is likely to arise from electron-beam lithography, which is already[66] providing line widths less than 1 μm. Technological reduction in the physical size of components will result in smaller parasitic capacitances and thereby allow[33] wider-bandwidth operational amplifiers for a given level of power dissipation and input bias current. This is likely to be of more consequence than ingenuity of internal circuit design in extending the upper frequency limit. The increasing use of 'quad' operational amplifiers reduces chip counts while making additional amplifiers available for summing and buffering purposes. They are also programmable by a single resistor feeding all amplifiers in parallel. However, denser packing causes inter-amplifier capacitive coupling[67] which can degrade filter performance. Other potentially attractive general-purpose analogue devices such as the DVCCS of Fig. 5.13 are only held back by processing difficulties which may soon be overcome. The onus is thus very much on improved fabrication techniques to extend[68] the capability of active filters to cover critical applications in the frequency range up to hundreds of kHz and where unusual specifications of low noise, zero offset, and wide dynamic and temperature ranges have presently to be met by passive networks.

But advances in technology also herald an increasing challenge from digital and sampled-data filters using charge-coupled or switched-MOS techniques. The progressive digitization of communication channels suggests carrying out filtering operations directly on the digital signals. Coding and filtering systems can then be compatibly integrated on a single chip. Hitherto, the hardware to implement the necessary delay units, multipliers, and summers has not been available for volume use, but the situation is changing[69] as industry masters large-scale integration. The frequency characteristics of digital filters are determined entirely by the coefficients supplied and hence the systems are truly universal and programmable. Digital specification assures stability, reliability, and repeatability as well as ease of testing. Sensitivity is also inherently low, but there are rounding errors and the method can only be used effectively in appropriate situations. Whenever the signal is already in analogue form, digital filtering is unnecessarily extravagant requiring as many as four delay elements and four multipliers merely to realize a second-order bandpass function. Furthermore, demands on sampling rates become excessive for higher-frequency filtering.

Other possible alternatives to active filters include electrothermal circuits[70] (ETC's) exploiting interactions between the thermal and electronic properties of devices, special solid-state functional blocks, and phase-locked loops (see Sec. 9.5.3). However, none of these can compete in variety of application with the RC-active filter, which, notwithstanding digital rivalry, is likely to continue its penetration of the communications, instrumentation, and other markets.

References

1. IEEE Standard Definitions, 'Glossary of terms in integrated electronics', *Bell Labs. Record*, p. 296, Oct.–Nov. 1966.
2. A. L. Grebene, *Analog Integrated Circuit Design*, Van Nostrand Reinhold, New York, 1972, Chap. 1.
3. C. A. Harper, *Handbook of Components for Electronics*, McGraw-Hill, New York, 1977.
4. Z. H. Meiksin, *Thin and Thick Films for Hybrid Microelectronics*, Lexington Books, Lexington, Mass., 1976, Chaps. 1, 9, and 10.
5. W. Worobey and J. Rutkiewicz, 'Tantalum thin-film RC circuit technology for a universal active filter', *IEEE Trans.*, **PHP-12**, 276–282, 1976.
6. Bell Telephone Laboratories, *The Physical Design of Electronic Systems*, Vol. III, *Integrated Devices and Construction Technology*, Prentice Hall, Englewood Cliffs, N.J., 1971.
7. P. J. Holmes and R. G. Loasby, *Handbook of Thick Film Technology*, Electrochemical Publications, London, 1976.
8. W. R. Bratschun, 'Glass-passivated thick-film capacitors for RC circuits', *IEEE Trans.*, **PHP-12**, 194–200, 1976.
9. G. S. Moschytz, *Linear Integrated Networks–Design*, Van Nostrand Reinhold, New York, 1974.
10. J. Loughran and K. Kurzweil, 'Economic considerations in multilayer thick-film hybrids', *IEEE Trans.*, **PHP-10**, 120–131, 1974.
11. M. I. Cohen, B. A. Unger, and J. F. Milkosky, 'Laser machining of thin films and integrated circuits', *Bell Systems Tech. J.*, **47**, 385–405, 1968.
12. C. L. Mohr, J. A. Ringo, E. H. Stevens, and R. A. Baker, 'Fundamental limitations on the laser trimming of active networks', *Proc. IEEE*, **65** (2), 269–270, 1977.

13. R. L. Lenington, 'Financial consciousness in the manufacturing of thin-film hybrids', *IEEE Trans.*, **MFT-5** (1), 13–17, 1976.
14. P. L. Kirby, 'Thick-film advances simplify complex hybrid module design', *Electron. Engng*, **48** (577), 35–38, 1976.
15. L. T. Bruton and A. B. Haase, 'Sensitivity of generalised immittance converter-embedded ladder structures', *Trans. IEEE*, **CAS-21**, 245–249, 1974.
16. E. A. Faulkner and J. B. Grimbleby, 'The effect of amplifier gain-bandwidth product on the performance of active filters', *Radio Electron. Engr*, **45** (9), 547–551, 1973.
17. G. Wilson, Y. Bedri, and P. Bowron, 'RC-active networks with reduced sensitivity to amplifier gain-bandwidth product', *IEEE Trans.*, **CAS-21** (5), 618–624, 1974.
18. R. Tarmy and M. S. Ghausi, 'Very high-Q insensitive active RC networks', *IEEE Trans.*, **CT-17** (4), 358–366, 1970.
19. C. F. Ho and P. L. Chiu, 'Realization of active-R filters using the amplifier pole', *Proc. IEE*, **123** (5), 406–410, 1976.
20. A. K. Mitra and V. K. Aatre, 'Low-sensitivity high-frequency active-R filters', *IEEE Trans.*, **CAS-23** (11), 670–676, 1976.
21. J. R. Brand and R. Schaumann, 'Active R filters: review of theory and practice', *IEEE Journal*, **ECS-2**, (4), 89–101, 1978.
22. L. T. Bruton and D. H. Treleaven, 'Electrical noise in lowpass FDNR filters', *IEEE Trans.*, **CT-20**, 154–161, 1973.
23. R. E. Thomas and D. J. Ray, 'High-frequency nonlinearities in active filters', *Proc. NEC*, **27**, 199–201, 1972.
24. J. G. Graeme, G. E. Tobey, and L. P. Huelsman, *Operational Amplifiers, Design and Applications*, McGraw-Hill, New York, 1971.
25. S. M. Shinners, *Modern Control System Theory and Applications*, Addison Wesley, Reading, Mass., 1971, Chap. 8.
26. A. Forsén and L. Kristiansson, 'Analysis of nonlinear model for operational amplifiers in active RC networks', *Int. J. Circuit Theory Applics*, **2**, 13–22, 1974.
27. K. Schlotzhauer and K. Singhal, 'Influence of capacitor losses on the Q of active filters', *Proc. IEEE*, **61**, 1662–1663, 1973.
28. H. C. Lin, *Integrated Electronics*, Holden-Day, San Francisco, 1967.
29. G. S. Moschytz, 'The operational amplifier in linear active networks', *IEEE Spectrum*, **7** (1), 42–47, 1970.
30. W. E. Hearn, 'Fast-slewing monolithic operational amplifier', *IEEE Journal*, **SC-6**, 20–24, 1971.
31. S. A. Boctor, 'Design of a 3rd-order single amplifier filter', *IEEE Trans.*, **CAS-22**, 329–334, 1975.
32. M. Baumwolspiner, 'Stability considerations in nonlinear feedback structures as applied to active networks', *Bell Systems Tech. J.*, **51** (9), 2029–2063, 1972.
33. P. R. Gray and R. G. Mayer, 'Recent advances in monolithic operational amplifier design', *IEEE Trans.*, **CAS-21**, 317–327, 1974.
34. R. W. Newcomb, *Active Integrated Circuit Synthesis*, Prentice Hall, Englewood Cliffs, N.J., 1968.
35. W. E. Heinlein and W. H. Holmes, *Active Filters for Integrated Circuits*, Prentice Hall, Englewood Cliffs, N.J., R. Oldenburg Verlag, Munich, 1974, Chap. 4.

36. D. R. Means, 'Compensation for temperature dependence of operational amplifier gain', *IEEE Journal*, **SC-7**, 507–509, 1972.
37. A. A. Gaash, R. S. Pepper, and D. O. Pederson, 'Design of integrable desensitised frequency selective amplifier', *IEEE Journal*, **SC-1** (1), 29–35, 1966.
38. M. Biey and A. Premoli, 'High-Q low-cost RC active filters: the RIMF section', *IEEE Trans*, **CAS-24**, 35–41, 1977.
39. A. S. Sedra and J. L. Espinosa, 'Sensitivity and frequency limitations of biquadratic active filters', *IEEE Trans.*, **CAS-22** (2), 122–130, 1975.
40. G. S. Moschytz, 'Gain-sensitivity product—a figure of merit for hybrid integrated filters using single operational amplifiers', *IEEE Journal*, **SC-6** (3), 103–110, June 1971.
41. A. L. Rosenblum and M. S. Ghausi, 'Multiparameter sensitivity in active RC networks', *IEEE Trans.*, **CT-18** (6), 592–599, 1971.
42. L. P. Huelsman, 'The distributed-lumped-active network: its application to filtering problems', *IEEE Spectrum*, **6** (8), 51–58, 1969.
43. B. A. Shenoi, 'Optimum variability design and comparative evaluation of thin-film RC-active filters', *IEEE Trans.*, **CAS-21** (2), 263–267, 1974.
44. K. Mossberg and D. Åkerburg, 'Accurate trimming of active RC filters by means of phase measurements', *Electron. Letters*, **5** (21), 520–521, 1969.
45. G. S. Moschytz, 'A general approach to twin-T design and its application to hybrid integrated linear active networks', *Bell Systems Tech. J.*, **49**, 1105–1149, 1970.
46. E. Lueder and G. Malek, 'Measure-predict tuning of hybrid thin-film filters', *IEEE Trans.*, **CAS-23** (7), 461–466, 1976.
47. R. A. Friedenson, R. W. Daniels, R. J. Dow, and P. H. McDonald, 'RC active filters for the D3 channel bank filter', *Bell Systems Tech. J.*, **54** (3), 507–529, 1975.
48. A. G. Albin and E. J. Swenson, 'Laser resistance trimming from the measurement point of view', *IEEE Trans.*, **PHP-8**, 14–19, 1972.
49. P. V. Lopresti, 'Optimum design of linear tuning algorithms', *IEEE Trans.*, **CAS-24** (3), 144–151, 1977.
50. G. J. Deboo and R. C. Hedlund, 'Automatically tuned filter uses I.C. operational amplifiers', *Electrical Design News*, **17**, 38–41, 1 February 1972
51. A. B. Grebene and H. R. Camenzind, 'Frequency-selective integrated circuits using phase-locked techniques', *IEEE Journal*, **SC-4** (4), 216–225, 1969.
52. E. Lueder, 'A decomposition of a transfer function minimising distortion and inband losses', *Bell Systems Tech. J.*, **49** (3), 455–469, 1970.
53. G. S. Moschytz, 'A second-order pole-zero pair selection for nth-order minimum sensitivity networks', *IEEE Trans.*, **CT-17** (4), 527–534, 1970.
54. K. R. Laker and M. S. Ghausi, 'Synthesis of a low-sensitivity multiloop feedback active RC filter', *IEEE Trans.*, **CAS-21** (2), 252–259, 1974.
55. J. Tow, 'Some results on generalised follow-the-leader feedback active filters', in *Proceedings of the IEEE International Symposium on Circuits and Systems, Phoenix, Arizona, April 25–27, 1977*, IEEE Publications, 1977, pp. 462–465.
56. J. Tow, 'Design and evaluation of shifted-companion form of active filters', *Bell Systems Tech. J.*, **54** (3), 545–568, 1975.

57. G. Hurtig, 'The primary resonator block technique of filter synthesis', in *IEEE International Filter Symposium, Santa Monica, California, April 1972*, IEEE Publications, 1972, p. 84.
58. D. J. Perry, 'New multiple-feedback active RC circuit', *Electron. Letters*, **11** (16), 364–365, 1975.
59. K. Martin and A. S. Sedra, 'Designing leapfrog and SFG filters with optimum dynamic range', *Proc. IEEE*, **65** (8), 1210–1211, 1977.
60. G. S. Moschytz, *Linear Integrated Networks: Fundamentals*, Van Nostrand Reinhold, New York, 1974, p. 224.
61. K. R. Laker and M. S. Ghausi, 'A comparison of active multiple-loop feedback techniques for realizing high-order bandpass filters', *IEEE Trans.*, **CAS-21** (6), 774–783, November 1974.
62. R. N. Gadenz, 'On low-sensitivity realizations of band elimination active filters', *IEEE Trans.*, **CAS-24** (4), 175–183, 1977.
63. L. B. Jackson, J. F. Kaiser, and H. S. McDonald, 'An approach to the implementation of digital filters', *IEEE Trans.*, **AU-16** (3), 413 419, 1968.
64. Frequency Devices Inc., 'Active filter', U.S. Patent No. 3,987,370, October 1976.
65. L. Nattera, 'Advances in film extend component capabilities', Preview of Electronic Components Conference, *Electronics*, **50** (11), 114, 1977.
66. H. Ahmed, 'Electron-beam lithography for microcircuit fabrication', *Electron. Power*, **22** (7), 433–436, 1976.
67. G. T. Volpe and A. Grabel, 'The evolution of an active filter from specification to production', in *Proceedings of the IEEE International Symposium on Circuits and Systems, Phoenix, Arizona, April 25–27, 1977*, IEEE Publications, 1977, pp. 299–303.
68. M. J. Riezenman, 'Special report: active filters ride the crest of new technology', *Electronics*, **50** (11), 119–124, 1977.
69. G. Edwards, 'Digital filters can simplify signal processing', *Electron. Engng*, **48** (6), 53–55, 1976.
70. W. J. Louw, D. J. Hamilton, and W. J. Kerwin, 'Inductor-less, capacitor-less state-variable electrothermal filters', *IEEE Journal*, **SC-12** (4), 416–420, 1977.

Appendix 1. Normalization and denormalization

For the purposes of standardization, ease of comparison, and calculation, it is convenient to tabulate approximation functions and synthesize networks in the normalized form. As a result, unnaturally large values of inductor and capacitor (e.g., 1 H and 1 F) and small values of resistor (e.g., a few ohms) occur in examples. A filter so designed for unity cut-off frequency and resistance termination must then be converted into terms of a more practical specification by processes of denormalization.

Frequency denormalization is carried out in terms of the normalized frequency variable $s_n = s/\omega_n$, where ω_n is the frequency denormalization factor, usually selected as the desired cut-off frequency ω_0 in radians per second. In order that the impedance of an element remains invariant, it is necessary for an inductance that

$$s_n L_n = sL = \omega_n s_n L \quad \text{or} \quad L = L_n/\omega_n$$

and for a capacitance that

$$1/s_n C_n = 1/sC = 1/\omega_n s_n C \quad \text{or} \quad C = C_n/\omega_n$$

where L, C represent denormalized and L_n, C_n represent normalized element values. Resistors, being independent of frequency, are unaffected by the process.

Impedance denormalization converts a normalized impedance to a practical level so as to avoid loading effects and minimize interaction with amplifier parasitics in active filters. The impedance denormalization factor Z_n is often taken as matching resistance R_0. In order that the element impedances are invariant,

$$sL = Z_n sL \quad \text{or} \quad L = Z_n L_n$$
$$1/sC = Z_n/sC_n \quad \text{or} \quad C = C_n/Z_n$$

and

$$R = Z_n R_n$$

where R is the denormalized and R_n the normalized value.

Combined denormalization carries out both scalings simultaneously. These are now listed for reference and include the FDNR of value D_n whose derivation from Sec. 8.3.2 is left as an exercise for the reader.

Denormalized element	L	C	R	D
Denormalized expression	$Z_n L_n / \omega_n$	$C_n / Z_n \omega_n$	$Z_n R_n$	$D_n / Z_n \omega_n^2$

Appendix 2. Predistortion

As explained in Sec. 9.2, the presence of component imperfections causes the performance of a practical filter to deviate from the theoretical. These effects may be corrected at the design stage by a technique known as *predistortion* which, essentially, pre-shifts the approximating function singularities in the opposite sense so that the required response is obtained.

In *LCR* filters, the dominant reactive losses can be represented by a series resistor R_s with each inductor (see Fig. 4.2) and a shunt conductor G_p with each capacitor (see Fig. 9.8). The respective element impedances are then

$$Z_L = sL + R_s = (s + \omega_1)L = pL$$

and

$$Z_c = [sC + G_p]^{-1} = [(s + \omega_1)C]^{-1} = [pC]^{-1}$$

where $\omega_1 = R_s/L = G_p/C$, as defined in Fig. 9.9, and the new frequency variable $p = s + \omega_1$.

Hence the impedances of the *RL* and *RC* combinations are the same functions of p as L or C are of s, the original leftward shift of the pole-zero pattern being countered by moving a distance ω_1 to the right. This obviously limits the amount of predistortion as the pole cannot cross the imaginary axis. The realization procedure for an impedance

$$Z_{12} = \frac{1}{s^2 + bs + 1}$$

is to synthesize the transformed function in the p plane

$$Z_{12}(p - \omega_1) = Z'_{12}(p) = \frac{1}{(p - \omega_1)^2 + b(p - \omega_1) + 1}$$

then add the dissipation ω_1 to the realized network to cancel the original predistortion. This does, however, require additional components and introduces unnecessary power loss. The technique can also be used to incorporate additional reactive elements.

In active networks, the effect of finite amplifier gain-bandwidth product produces incremental shifts in selectivity and centre frequency of a resonator section so that

$$\hat{Q} = Q(1 + \delta Q) = Q + \Delta Q$$

and

$$\hat{\omega}_0 = \omega_0(1 + \delta\omega_0) = \omega_0 + \Delta\omega_0$$

by Eq. (9.5). These perturbations can be designed out by predistorting the prescribed transfer function with selectivity $(Q - \Delta Q)$ and resonant frequency $(\omega_0 - \Delta\omega_0)$ followed by formulation in terms of ideal amplifiers. In the active case, no additional elements are introduced whilst the effective pre-shift in pole locations is to the left, i.e., to a more stable position.

References

1. S. Darlington: 'Synthesis of reactance 4-poles which produce prescribed insertion loss characteristics', *J. Math. Phys.*, **18**, 257–353, 1939.
2. N. Balabanian: *Network Synthesis*, Prentice-Hall, Englewood Cliffs, N.J., 1958, Chap. 9.

Appendix 3. Nodal admittance matrix

Nodal analysis of a general network yields the matrix equation

$$[I] = [Y][V]$$

where $[I]$, $[V]$ are column vectors and $[Y]$ is the (definite) nodal admittance matrix (NAM), assuming that a reference node has been chosen (essential if a non-trivial solution is to be obtained). $[Y]$ may be formed by inspection of the network diagram with the aid of the following definitions:

(a) Y_{ii} is the sum of all admittances attached to node i (i.e., the self-admittance of node i);
(b) Y_{ij} is the negative sum of all admittances connected directly between nodes i and j (i.e., the mutual admittance between nodes i and j).

In a passive network, $Y_{ij} = Y_{ji}$ ($i \neq j$) and the off-diagonal terms are symmetrically distributed about the leading diagonal.

Active devices are accommodated simply by adding the active matrix elements at the appropriate node positions. Thus, in Fig. A3.1, the VCVS matrix is added between nodes 3 and 4. The presence of active elements disturbs the symmetry of the NAM, as might be expected by inspecting the admittance matrices of the various devices considered in Chapter 5.

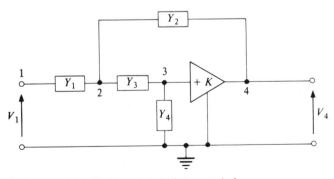

Fig. A3.1. Active network labelled for nodal-admittance analysis

Although network functions may be derived from the NAM by means of Cramer's rule, many computer analysis programs have utilized pivotal condensation, which systematically reduces an $[n \times n]$ matrix about any desired node. Thus, if any node i in an $[n \times n]$ matrix is to be suppressed, terms in the resulting $[n-1] \times [n-1]$ matrix are given by

$$Y'_{rs} = Y_{rs} - \frac{Y_{ri} Y_{is}}{Y_{ii}}$$

where Y'_{rs} is the modified version of Y_{rs}. This process is repeated about successive nodes until the required degree of reduction is obtained. Usually, the point of completion is the $[2 \times 2]$ matrix, although occasions do arise when some lesser degree of reduction is necessary. The mechanics of the analysis are illustrated by application to the circuit shown in Fig. A3.1 for which the NAM is

$$[Y] = \begin{array}{c} \\ 1 \\ 2 \\ 3 \\ 4 \end{array} \begin{array}{|cccc|} \hline 1 & 2 & 3 & 4 \\ \hline Y_1 & -Y_1 & 0 & 0 \\ -Y_1 & Y_1 + Y_2 + Y_3 & -Y_3 & -Y_2 \\ 0 & -Y_3 & Y_3 + Y_4 & 0 \\ 0 & -Y_2 & -KY'_{44} & Y_2 + Y'_{44} \\ \hline \end{array} \quad (A3.1)$$

where Y'_{44} refers to the VCVS and $Y'_{33} = Y'_{34} = 0$. To obtain the voltage transfer function V_4/V_1, it is necessary to eliminate nodes 2 and 3. Elimination of node 3 yields a $[3 \times 3]$ matrix

$$[Y] = \begin{array}{c} \\ 1 \\ 2 \\ 4 \end{array} \begin{array}{|ccc|} \hline 1 & 2 & 4 \\ \hline Y_1 & -Y_1 & 0 \\ -Y_1 & Y_1 + Y_2 + Y_3 - Y_3^2/(Y_3 + Y_4) & -Y_2 \\ 0 & -Y_2 - Y_3 K Y'_{44}/(Y_3 + Y_4) & Y_2 + Y'_{44} \\ \hline \end{array}$$

Pivotal condensation about node 2 gives

$$\frac{V_4}{V_1} = -\frac{Y_{41}}{Y_{44}}$$

$$= \frac{Y_1 \left(-Y_2 + \dfrac{Y_3 K Y'_{44}}{(Y_3 + Y_4)}\right) \Big/ \left\{Y_1 + Y_2 + Y_3 - \dfrac{Y_3^2}{Y_3 + Y_4}\right\}}{\left[Y_2 + Y'_{44} - Y_2 \left\{Y_2 + \dfrac{Y_3 K Y'_{44}}{(Y_3 + Y_4)}\right\} \Big/ \left\{Y_1 + Y_2 + Y_3 - \dfrac{Y_3^2}{Y_3 + Y_4}\right\}\right]}$$

(A3.2)

APPENDIX 3. NODAL ADMITTANCE MATRIX

For an ideal VCVS, $Y'_{44} \to \infty$ and the above expression reduces to

$$\frac{V_4}{V_1} = \frac{Y_1 Y_3 K}{(Y_1 + Y_2 + Y_3)(Y_3 + Y_4) - Y_3^2 - Y_2 Y_3 K} \tag{A3.3}$$

This result is obtained more directly if an ideal active device is assumed at the outset. The VCVS introduces[1] a constraint in the network since node 4 is forced to follow node 3 ($V_4 = KV_3$) without loading node 3. Hence, the constrained matrix is formed directly from the unconstrained (passive) matrix by first adding $K \times$ (column 4) to column 3, then deleting row and column 4. In the example, the constrained matrix is formed from Eq. (A3.1) as

$$[Y] = \begin{array}{c} \\ 1 \\ 2 \\ 3 \end{array} \begin{array}{|c|c|c|} \hline 1 & 2 & 3 \\ \hline Y_1 & -Y_1 & 0 \\ \hline -Y_1 & Y_1 + Y_2 + Y_3 & -Y_3 - KY_2 \\ \hline 0 & -Y_3 & Y_3 + Y_4 \\ \hline \end{array}$$

Condensation about node 2 yields V_3/V_1, from which V_4/V_1 is obtained by noting that $V_4 = KV_3$. The result confirms that of Eq. (A3.3) but obviates the limiting process and lengthier expressions associated with Eq. (A3.2).

The technique may be extended to operational devices, the constraining equations and operations to form the matrix being:

(a) *Single-input OA*. $V_k = AV_i$ so that, as $A \to \infty$, $V_i (= V_k/A) \to 0$. Thus, delete column i and row k.

(b) *Differential-input OA*. $V_k = A(V_i - V_j)$ so that, as $A \to \infty$, $V_i \to V_j$. Thus, add column i to column j and delete either column i or j, together with row k.

Evaluation of the voltage transfer function (V_0/V_I) in each of the operational cases is most conveniently obtained as

$$\frac{V_0}{V_I} = \frac{|Y_{10}|}{|Y_{11}|}$$

where $|Y_{10}|$ and $|Y_{11}|$ are cofactors of the constrained admittance matrix while I and 0 refer to input and output nodes, respectively.

Reference

1. A. Nathan, 'Matrix analysis of networks having infinite-gain operational amplifiers', *Proc. IRE*, **49** (10), 1577–1578, 1961.

Appendix 4. Signal-flow graphs

The signal-flow graph is a topological representation of a set of linear equations such as those arising from nodal analysis. For a network containing VCVSs, the flow graph is constructed as follows:

(a) Number each node (except the reference) in the active network and assign a corresponding node in the flow graph.

(b) For a VCVS of gain K_{pq} occurring between nodes p and q, connect corresponding nodes in the flow graph. Associate a gain K_{pq} with the resulting branch and direct it *towards* node q (the VCVS output node).

(c) At each node (except those to which are connected ideal voltage sources) assign *incoming* branches of gain (Y_{ij}/Y_{jj}), where Y_{ij} is (-1) × (mutual admittance between nodes i and j), Y_{jj} is the self-admittance of node j, and j is the node under consideration. Since an ideal voltage source has $Y_{jj} = \infty$, it can *receive* no branch. For this reason, the signal source has only *outgoing* branches and VCVS outputs receive only a single branch, corresponding to the VCVS gain.

The resulting flow graph for the active network of Fig. A3.1 is shown in Fig. A4.1. The derivation of transfer functions from signal-flow graphs is ameliorated by use of Mason's general rule[1] for the gain between nodes A and B as

$$G_{AB} = \frac{1}{\Delta} \cdot \sum_k G_k \Delta_k$$

where G_k is the gain of the kth forward path between nodes A and B, Δ is the graph determinant, calculated as

$1 - \sum$ (single-feedback path gains)
$+ \sum$ (products of two non-touching feedback-path gains)
$- \sum$ (products of three non-touching feedback-path gains)
$+$ etc.

and Δ_k is the value of Δ with all terms touching the kth forward path omitted. In this context, 'non-touching' is taken to mean 'having no nodes in common with'.

APPENDIX 4. SIGNAL-FLOW GRAPHS

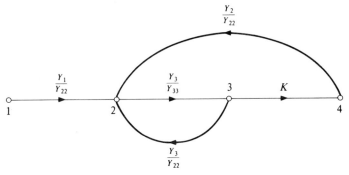

Fig. A4.1. Signal-flow graph for active network of Appendix 3

Applying Mason's rule to evaluate the voltage transfer ratio V_4/V_1 from Fig. A4.1 gives

$$\Delta = 1 - \left(\frac{Y_3^2}{Y_{22}Y_{33}} + \frac{KY_2Y_3}{Y_{22}Y_{33}}\right)$$

$$G_1 = \frac{KY_1Y_3}{Y_{22}Y_{33}}, \quad \text{and} \quad \Delta_1 = 1.$$

Hence

$$\frac{V_4}{V_1} = \frac{KY_1Y_3}{Y_{22}Y_{33} - Y_3^2 - KY_2Y_3}$$

which, after simplification, yields the same result as that derived in Appendix 3.

Reference

1. S. J. Mason: 'Feedback theory—further properties of signal-flow graphs', *Proc. IRE*, **44**, 920–926, 1956.

Appendix 5. Matrix conversion table

From \ To	[y]	[a]	[z]	[h]	[g]
[y]		$\dfrac{-y_{22}}{y_{21}}\quad \dfrac{-1}{y_{21}}$ $\dfrac{-\|y\|}{y_{21}}\quad \dfrac{-y_{11}}{y_{21}}$	$\dfrac{y_{22}}{\|y\|}\quad \dfrac{-y_{12}}{\|y\|}$ $\dfrac{-y_{21}}{\|y\|}\quad \dfrac{y_{11}}{\|y\|}$	$\dfrac{1}{y_{11}}\quad \dfrac{-y_{12}}{y_{11}}$ $\dfrac{y_{21}}{y_{11}}\quad \dfrac{\|y\|}{y_{11}}$	$\dfrac{\|y\|}{y_{22}}\quad \dfrac{y_{12}}{y_{22}}$ $\dfrac{-y_{12}}{y_{22}}\quad \dfrac{1}{y_{22}}$
[a]	$\dfrac{a_{22}}{a_{12}}\quad \dfrac{-\|a\|}{a_{12}}$ $\dfrac{-1}{a_{12}}\quad \dfrac{a_{11}}{a_{12}}$		$\dfrac{a_{11}}{a_{21}}\quad \dfrac{\|a\|}{a_{21}}$ $\dfrac{1}{a_{21}}\quad \dfrac{a_{22}}{a_{21}}$	$\dfrac{a_{12}}{a_{22}}\quad \dfrac{\|a\|}{a_{22}}$ $\dfrac{-1}{a_{22}}\quad \dfrac{a_{21}}{a_{22}}$	$\dfrac{a_{21}}{a_{11}}\quad \dfrac{-\|a\|}{a_{11}}$ $\dfrac{1}{a_{11}}\quad \dfrac{a_{12}}{a_{11}}$
[z]	$\dfrac{z_{22}}{\|z\|}\quad \dfrac{-z_{12}}{\|z\|}$ $\dfrac{-z_{21}}{\|z\|}\quad \dfrac{z_{11}}{\|z\|}$	$\dfrac{z_{11}}{z_{21}}\quad \dfrac{\|z\|}{z_{21}}$ $\dfrac{1}{z_{21}}\quad \dfrac{z_{22}}{z_{21}}$		$\dfrac{\|z\|}{z_{22}}\quad \dfrac{z_{12}}{z_{22}}$ $\dfrac{-z_{21}}{z_{22}}\quad \dfrac{1}{z_{22}}$	$\dfrac{1}{z_{11}}\quad \dfrac{-z_{12}}{z_{11}}$ $\dfrac{z_{21}}{z_{11}}\quad \dfrac{\|z\|}{z_{11}}$
[h]	$\dfrac{1}{h_{11}}\quad \dfrac{-h_{12}}{h_{11}}$ $\dfrac{h_{21}}{h_{11}}\quad \dfrac{\|h\|}{h_{11}}$	$\dfrac{-\|h\|}{h_{21}}\quad \dfrac{-h_{11}}{h_{21}}$ $\dfrac{-h_{22}}{h_{21}}\quad \dfrac{-1}{h_{21}}$	$\dfrac{\|h\|}{h_{22}}\quad \dfrac{h_{12}}{h_{22}}$ $\dfrac{-h_{21}}{h_{22}}\quad \dfrac{1}{h_{22}}$		$\dfrac{h_{22}}{\|h\|}\quad \dfrac{-h_{12}}{\|h\|}$ $\dfrac{-h_{21}}{\|h\|}\quad \dfrac{h_{11}}{\|h\|}$
[g]	$\dfrac{\|g\|}{g_{22}}\quad \dfrac{g_{12}}{g_{22}}$ $\dfrac{-g_{21}}{g_{22}}\quad \dfrac{1}{g_{22}}$	$\dfrac{1}{g_{21}}\quad \dfrac{g_{22}}{g_{21}}$ $\dfrac{g_{11}}{g_{21}}\quad \dfrac{\|g\|}{g_{21}}$	$\dfrac{1}{g_{11}}\quad \dfrac{-g_{12}}{g_{11}}$ $\dfrac{g_{21}}{g_{11}}\quad \dfrac{\|g\|}{g_{11}}$	$\dfrac{g_{22}}{\|g\|}\quad \dfrac{-g_{12}}{\|g\|}$ $\dfrac{-g_{21}}{\|g\|}\quad \dfrac{g_{11}}{\|g\|}$	

with determinants given by

$$|y| = y_{11}y_{22} - y_{12}y_{21}$$
$$|a| = a_{11}a_{22} - a_{12}a_{21}$$
$$|z| = z_{11}z_{22} - z_{12}z_{21}$$
$$|h| = h_{11}h_{22} - h_{12}h_{21}$$
$$|g| = g_{11}g_{22} - g_{12}g_{21}$$

Appendix 6. Sample design

The following example is an illustration of the design procedure for a filter from an initial response specification through to a final practical circuit suitable for construction in the laboratory or the production situation. While more stringent requirements can occur in practice, the general approach should serve as a useful guide, particularly to those unfamiliar with filter principles.

Specification

To design an RC-active filter to meet the symmetrical bandpass specification limits shown in Fig. A6.1.

Approximation

Since the entire specification cannot be met by a single second-order resonator section of arbitrary Q, resort is made to the standard approximating functions discussed in Chapter 2.

Comparing Fig. A6.1 with Fig. 2.7(b), it is seen that $\gamma_{min} = 18$ dB and $\gamma_{max} = 2$ dB. These are unchanged in the equivalent lowpass function but occur at different frequencies, determined with the aid of Table 2.3.

Noting that $f = \omega/2\pi$, the equivalent normalized lowpass stopband frequency (see Fig. 2.2.) is

$$\Omega'_2 = \frac{\omega'_2}{\omega_0} = \frac{f_4 - f_3}{f_2 - f_1} = \frac{2000 - 500}{1250 - 800} = \frac{10}{3}$$

and the corresponding normalized passband cut-off frequency $\Omega'_1 = \omega'_1/\omega_0 = 1$.

Trying first the simple maximally flat approximation, Eq. (2.7) gives the order of Butterworth function required to meet these specifications as

$$n_B = \frac{\log_{10}\left(\frac{10^{1.8} - 1}{10^{0.2} - 1}\right)}{2 \log_{10}(10/3)} = 1.94$$

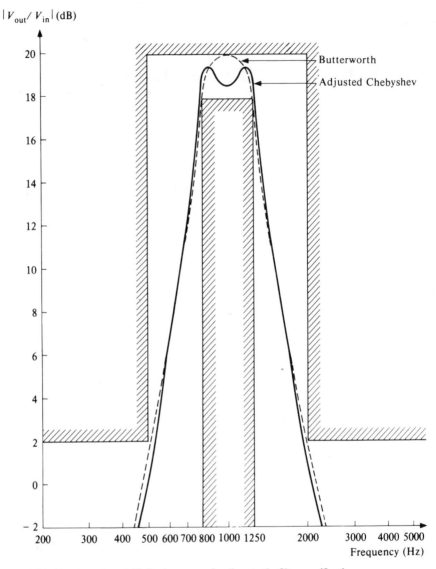

Fig. A6.1. Butterworth and Chebyshev approximations to the filter specification

This indicates that a second-order function is required which, from Table 2.1, has the normalized lowpass form

$$T(p) = \frac{H}{p^2 + 1.414\,214\,p + 1} \tag{A6.1}$$

Note, however, that this function exhibits a 3 dB loss at the edge of the passband. Since γ_{max} is 2 dB, Eq. (2.4) must be used to determine

$$\varepsilon = \sqrt{10^{\gamma_{max}/10} - 1} = 0.764\,783.$$

APPENDIX 6. SAMPLE DESIGN

From Eq. (2.3), it is now necessary to transform the frequency variable so that $\Omega' = \varepsilon^{1/n}\Omega$. This is accomplished by replacing p in Eq. (A6.1) by

$$\varepsilon^{1/2}p = 0.874\,519\,p$$

yielding the normalized lowpass form as

$$T(p) = \frac{H}{0.764\,783\,p^2 + 1.236\,757\,p + 1} \qquad (A6.2)$$

which has a loss of 2 dB at $\Omega = 1$ cut-off, as required by $\gamma_{max} = 2$ dB.

Using Eq. (2.23), transform $T(p)$ back to the normalized ($\omega_0 = 1$) bandpass form, so that, from Table 2.3,

$$p = Q\left(s + \frac{1}{s}\right) = \frac{s^2 + 1}{0.45s} \qquad (A6.3)$$

leads to the final normalized form of transfer function

$$T(s) = \frac{2.647\,810s^2}{s^4 + 0.727\,711\,s^3 + 2.264\,781s^2 + 0.727\,711s + 1} \qquad (A6.4)$$

where $H = 10$ so as to ensure a midband gain of 20 dB. Equation (A6.4) satisfies the stipulated bandpass specifications as indicated by the computer plot shown dashed in Fig. A6.1 which, for comparative purposes, has been denormalized to a centre-band frequency of 1000 Hz. However, this leaves no tolerance margin at the passband edges, small component imperfections and changes such as described in Sec. 9.2 being sufficient to cause the characteristic to move out of specification. Indeed, if the boundary lines of Fig. A6.1 are regarded as worst-case limits, good design technique suggests that the final approximation should not approach them too closely at any frequency.

Since significant improvement is not possible from modified forms of the Butterworth approximation, consideration is now given to the Chebyshev function of Eq. (2.8). In this case, for $\gamma_{max} = 1$ dB and $\gamma_{min} = 18$ dB, the passband and stopband clearances are equalized at 0.5 dB and Eq. (2.11) gives $n_c = 1.83$. Hence, a second-order version suffices and Table 2.1 gives the normalized lowpass function having 1 dB passband ripple as

$$T(p) = \frac{H}{p^2 + 1.097\,734\,p + 1.102\,510}$$

which, when transformed using Eq. (A6.3), results in a denormalized characteristic having passband edges at 789 Hz and 1268 Hz with stopband limits at 552 Hz and 1810 Hz. These represent respective clearances on the specification of 1.4 and 10 per cent, suggesting the possibility of equalization by modifying the effective Q in Eq. (A6.3) to a value of 2, thereby yielding the final normalized characteristic, centred about the midband-gain level of 18.5 dB, as

$$T(s) = \frac{2.319\ 121s^2}{s^4 + 0.548\ 867s^2 + 2.275\ 628s^2 + 0.548\ 867s + 1} \quad \text{(A6.5)}$$

The response is shown in Fig. A6.1 and has passband-edge clearances of 4 per cent (769 Hz and 1301 Hz) and stopband-edge clearances of 4.15 per cent (521 Hz and 1918 Hz).

Realization

For reasons stated in Secs. 4.6.2 and 9.6.1, $T(s)$ is preferably factored to allow realization as a cascade of two second-order bandpass functions, each expressed in the form of Eq. (6.63) as

$$T(s) = \frac{H_1 s}{[s^2 + (\omega_{01}/Q_1)s + \omega_{01}^2]} \cdot \frac{H_2 s}{[s^2 + (\omega_{02}/Q_2)s + \omega_{02}^2]} \quad \text{(A6.6)}$$

where $H_1 H_2 = 2.319\ 121$. Hence, with the aid of a readily available[1] computer library procedure,

$$T(s) = \underbrace{\frac{H_1 s}{(s^2 + 0.334\ 884s + 1.564\ 985)}}_{\text{Stage 1}} \cdot \underbrace{\frac{H_2 s}{(s^2 + 0.213\ 983s + 0.638\ 983)}}_{\text{Stage 2}}$$

(A6.7)

indicating equal selectivities $Q_1 = Q_2 = 3.7356$. In the absence of specific noise information, it is convenient to equalize the peak gains, thereby obviating the

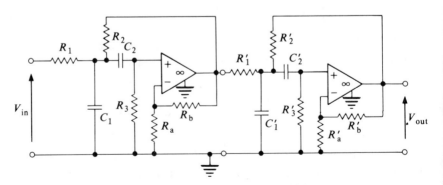

Fig. A6.2. Cascade realization using positive-gain controlled-source structure

APPENDIX 6. SAMPLE DESIGN

need in this case for a set order of cascade. This requires

$$\frac{H_1 Q_1}{\omega_{01}} = \frac{H_2 Q_2}{\omega_{02}} \quad \text{or} \quad \frac{H_1}{0.334\,884} = \frac{H_2}{0.213\,983}$$

which, since $H_1 H_2 = 2.319\,121$, leads to $H_1 = 1.905\,107$ and $H_2 = 1.217\,318$.

A suitable practical realization of $T(s)$ is afforded by the positive-gain controlled-source arrangement of Fig. A6.2. Replacing conductances by resistors in Eq. (6.3) results in the following set of design equations for the first stage:

$$\frac{K}{R_1 C_1} = 1.905\,107 \qquad (A6.8a)$$

$$\frac{(1-K)}{R_2 C_1} + \frac{(R_1 + R_3)}{R_1 R_3 C_1} + \frac{1}{R_3 C_2} = 0.334\,884 \qquad (A6.8b)$$

$$\frac{(R_1 + R_2)}{R_1 R_2 R_3 C_1 C_2} = 1.564\,985 \qquad (A6.8c)$$

As there are six unknowns and only three equations, some preliminary choice of element values is necessary. Unfortunately, Eqs. (A6.8a–c) preclude the use of a voltage follower since $K = 1$ requires

$$\frac{1}{R_3 C_1} + \frac{1}{R_3 C_2} = -1.570\,223$$

which is clearly impossible. Since any other choice of K demands two special gain-setting resistors (see Table 5.4), it is advisable to make these equal by setting $K = 2$. Furthermore, for ease of fabrication, let normalized $C_1 = C_2 = 1$ F. Substitution in Eq. (A6.8a) gives $R_1 = 1.049\,810$ Ω and the two remaining resistors are obtained by solving the following simultaneous pair:

$$\frac{2}{R_3} - \frac{1}{R_2} = -0.617\,670 \qquad (A6.9a)$$

$$\frac{1}{R_3}\left(\frac{1}{R_2} + 0.952\,554\right) = 1.564\,985 \qquad (A6.9b)$$

to yield $R_2 = 0.565\,575$ Ω and $R_3 = 1.738\,462$ Ω.

An exactly similar procedure applies to the second stage, for which the normalized design values are $K' = 2$, $C_1' = C_2' = 1$ F, $R_1' = 1.642\,956$ Ω, $R_2' = 0.885\,118$ Ω, and $R_3' = 2.720\,655$ Ω.

Denormalization

In order to convert the final design into terms of the original practical specifications, a frequency denormalization factor of $\omega_n = 2\pi f_0 = 2\pi \times 10^3$ is required. Impedance denormalization necessitates consideration of the non-ideal terminating resistances R_{in} and R_{out} of the amplifier together with the range of impedance level practicable in a given technology. For minimum interaction, the lowest resistor (R_2) must be significantly greater than the OA output resistance R_{out}, i.e., for the popular µA741 in Table 9.1, $0.565\,575R \gg 75\,\Omega$ while the highest resistor (R_3') must be significantly less than the OA input resistance R_{in}, i.e., $2.720\,655\,R \ll 5 \times 10^5\,\Omega$. Hence, the resistance level R or denormalization factor Z_n may be estimated from the geometric mean:

$$Z_n = R = \sqrt{\frac{5 \times 10^5 \times 75}{2.720\,655 \times 0.565\,575}} = 4.937\,\text{k}\Omega$$

The associated problem of deciding the gain-defining resistors R_a and R_b (constituting Y_a and Y_b in Table 5.4) is similarly resolved as

$$R_a = R_b = \sqrt{R_{out} R_{in}} = 6.124\,\text{k}\Omega.$$

For production runs using discrete components, it may be possible to optimize the choice of Z_n such that all, or at least several, of the resistors fall within an acceptable tolerance of preferred values. Such an exercise is not attempted here, nor is it necessary if laser-trimmed hybrid integrated-circuit production (Sec. 9.1.6) is contemplated. In the latter case, substrate area (Sec. 9.4.3) requirements are of concern, restricting the size of the resistors.

Application of ω_n and Z_n as in Appendix 1 yields the final practical component values for Fig. A6.2 as listed in Table A6.1. The operational amplifier, of course, needs to be energized with stabilized direct voltages of some ± 20 V (according to type).

Obviously, these values are feasible for hybrid integrated fabrication. In the case of restrictions, greater freedom of design can be obtained by replacing the µA741's by improved operational amplifiers. The above procedure suggests

Table A6.1 Final practical values for components in Fig. A6.2

Stage 1	Stage 2
$C_1 = C_2 = 32.24$ nF	$C_1' = C_2' = 32.24$ nF
$R_1 \quad = 5.183$ kΩ	$R_1' \quad = 8.111$ kΩ
$R_2 \quad = 2.792$ kΩ	$R_2' \quad = 4.370$ kΩ
$R_3 \quad = 8.583$ kΩ	$R_3' \quad = 13.432$ kΩ
$R_a = R_b = 6.124$ kΩ	$R_a' = R_b' = 6.124$ kΩ

APPENDIX 6. SAMPLE DESIGN

itself as a possible application for CAD techniques, the complete design being performed by an all-purpose computer program.

Alternative realizations

Other numerical examples are included in the main body of the text, as listed in Table A6.2. For moderately demanding filter specifications as in the above example, the reader will find the positive-gain controlled source and/or multiple-loop feedback circuits adequate. Lowpass, bandpass, and highpass realizations incorporating these structures are facilitated by use of Tables 6.1 and 6.2 and Eq. 6.3 for the former and Tables 7.1 and 7.2 for the latter.

Table A6.2   Location and nature of numerical design examples in text

Example no.	Page no.	Network structure	Function realized
6.1	*131*	Single-source positive-gain	Lowpass (medium Q)
6.2	*136*	Single-source positive-gain	Notch (medium Q)
6.4	*150*	Mixed-source unity-gain	Bandpass (medium/high Q)
6.5	*154*	FEN	Bandpass (medium/high Q)
7.1	*164*	Single-input single-feedback (bridged-T)	Bandpass (low Q)
7.2	*171*	Multiple-loop feedback	Lowpass (low Q)
7.3	*175*	General differential input	Bandpass (medium/high Q)
8.2	*196*	Leapfrog	Lowpass (fifth order)
8.3	*201*	State-variable	Bandpass (high Q)
8.4	*205*	Inductive subnetwork	Lowpass (fifth order)
8.5	*207*	FDNR	Lowpass (fifth order)
8.6	*210*	Embedded FDNR	Bandpass (sixth order)

Reference

1. K. W. Ellenberger, 'Programming the numerical solution of polynomial equations', *Commun.*, **ACM3**, 644–647, 1960.

Index

Active isolation, 138
Active ladder structures, 191–193
Active-R filters, 233–235, 244
Allpass filters, 176–177, 179, 231, 244
Allpass-pair circuit, 231, 232, 244, 246
Allpass response, 21, 128, 177
All-pole function, 169, 179
Amplifier (*see* Controlled sources; Current, Operational *and* Voltage amplifiers)
Analogue methods, 193–202, 213
Analytic continuation, 11
Applications of active filters, 2–4
   communications, 2–3, 20, 249, 257, 259
   entertainment electronics, 4, 243
   instrumentation, 1, 2, 3–4, 245, 246, 250, 259
   medical electronics, 3–4
   ultrasonics, 243
   waveform correction, 177
Approximation, 5, 9–24, 273–276
   approximating functions, 12–13, 18
   delay, 19–21
   elliptic, 16–19, 197, 257
   equi-ripple, 14–16
   least-squares interpolation, 9
   magnitude, 10–19
   maximally flat, 11, 171, 256, 273–275
   monotonic, 11, 14

Balanced network, 36, 41, 163
Bandpass filters, 3, 127, 132–133, 146, 150–151, 154, 157, 158–159, 164, 170, 175, 179, 181, 183, 185, 191, 197, 201, 210–211, 214–215, 216, 235, 273–279
Bandpass response, 6, 21, 53, 58–59, 126, 145, 157, 160, 164, 175, 200, 211, 274
Bandstop filter, 133, 168, 191
Bandstop response, 21, 41, 56–57, 59
Bessel polynomial, 20, 256
Biquadratic function, 152
Bridged ladder, 45, 46
Bridged-T network, 43, 149, 165, 173
   in single-feedback circuit, 165, 246
   with difference input, 146–147
Bridged twin-T network, 167, 173
Butterworth polynomial (*see also* Approximation, maximally flat), 11

Calahan decomposition, 65–67, 150
Cascade realizations, 10, 68, 148, 193, 196, 276–279
   pole-zero pairing, 252
   positioning, 252
   sensitivity, 72, 252
Cauer ladder, 38
Centreband gain (K_0), 56, 243, 248, 251
   FEN, 154
   single-feedback bridged-T, 165
   state-variable filter, 201
   unity-gain mixed-source filter, 150
Characterizing function, 10, 14, 16
Chebyshev polynomial (*see also* Approximation, equi-ripple), 14
Circulator, 118–119, 122, 177, 192, 213, 216
   admittance matrix, 217
   multi-port, 190, 192
Coefficient-matching technique, 6, 41, 44, 128, 141, 170, 179, 253
Commercial availability, 6, 254–257
Components (*see also* Operational amplifier, Transistors):
   count, 49, 179, 239
   discrete, 223–225, 257
   distributed, 47–50
   hybrid, 221, 227
   imperfections (*see also* Operational amplifier), 54–55, 228–233
   monolithic, 2, 221, 225
   passive parasitics, 236–237, 244, 258
   spread, 6, 142, 160, 168, 182, 216, 246
   thick-film, 2, 161, 226–227
   thin-film, 2, 161, 168, 225–226
   tolerances, 216, 223–226
Computer-aided techniques, 5, 7, 216
   for approximation, 9, 17
   for filter design, 279
   for multiparameter-sensitivity minimisation, 80
   in ladder simulation, 216
Controlled positive feedback (*see also* RC-embedded structures, 134, 161
Controlled sources, 5, 89, 92, 123
   current-controlled current (CCCS), 93, 99, 148

INDEX

Controlled sources—*contd.*
  current-controlled voltage (CCVS), 93, 101, 153
  DVCCS, 103, 160, 258
  DVCCS/DVCVS, 104, 154–157, 160, 244
  multi-terminal, 102
  voltage-controlled current (VCCS) 93, 97–98, 113
  voltage-controlled voltage (VCVS), 93, 94–97, 125, 132–133, 139, 148, 197, 276
Controlled-source filters, 125–162, 243–244, 245, 256, 276–277, 279
Converter, impedance, 104
  current generalized (CGIC), 107, 116
  current negative (CNIC), 105, 111, 115
  generalized (GIC), 107, 115
  negative (NIC), 105, 109
  positive (PIC), 106
  voltage generalized (VGIC), 107
  voltage negative (VNIC), 106
Converters (*see* GIC, NIC, etc.):
  matric description, 106, 123
  multi-port, 117
Cost, 241
Cramer's rule, 4, 45, 268
Current amplifiers, 88, 93–94
  grounded unity grain (GCUGA), 148
Current conveyor, 100
Current mirror, 100–101

Dasher cascade synthesis, 41
Decomposition:
  bandpass positive-feedback, 72, 143
  difference, 64–65
  in ladder synthesis, 40
  optimal-sensitivity, 64–68
  prescribed-sensitivity, 67–68
  summation, 65–67
Delay:
  approximation, 19–21
  equalization, 21
Denormalization, 263–264, 278
Design:
  examples, 279
  limits in Q-frequency plane, 237–239
  procedure sample, 273–279
  tables, 158–159, 181, 214–215
Difference-input bridged-T network, 146, 147
Differential-input (generalized) configuration, 174–177, 179, 180, 244, 246
Differential-input single-feedback circuits, 172–174, 180
Differential voltage-controlled current source (DVCCS), 103, 258
Digital filters, 1, 6, 24, 254, 258–259
Digital transmission, 19
Distortion, 252
Distributed-lumped active networks (DLANs), 244
Distributed-lumped notch filter, 50

Distributed-parameter networks, 24, 47–50, 61, 227, 245
  Telegraphers' equations, 49
  thin-film implementation, 48, 225
Divisor polynomial, 126, 150, 175
Double-ladder (Rauch) structure, 169
Driving-point immittance functions, RC, 32, 33–34, 39, 40, 88, 126, 155, 175
Duality and inversions, 31, 132
Dual-source realizations, 138–147, 156, 160
Dynamic range, 235–236, 253, 254, 258

Electrothermal circuits, 259
Elliptic functions, 16–17, 257
Energy principles:
  definition of Q, 54, 81
  realizability, 5, 32–35
  summed sensitivity, 78, 81
Environmental effects, 6, 52, 59, 216, 220, 239–241
Equal-capacitor constraint, 157
Equivalent networks, 79
Euler's formula, 77
Extended node-introduction synthesis, 172

FDNR (*see* Frequency-dependent negative resistance)
Feedback:
  allpass, 72
  and sensitivity, 62
  bandpass positive (RC-embedded), 72, 143–147, 160, 239
  compensation, 239
  in realizing controlled sources, 95–97, 101–103
  multiple-loop (FLF), 253–254
  notch-function negative, 72
Feedforward; multiple-loop FLF, 253–254
FEN (*see* Frequency-emphasizing network)
Fialkow-Gerst realizability conditions, 37
  tree structure, 41
Figure of merit, 6, 246
Filter, 1
  fixed-frequency, 256
  manufacturers of, 201, 255
  multichannel, 256–257
  RC-active, 1
  special-purpose, 257
  tunable, 256
  types, 1
Floating inductor simulation, 189–192
  parasitic elements, 217
Foster connection, 38
Frequency:
  centre (ω_0), 53, 55, 166, 238
  denormalization, 263–264, 278
  limits of operation, 6, 243–245, 258
  perturbation, 231–232
  tracking, 250–251
  transformations, 21–24

INDEX

Frequency-dependent negative resistance (FDNR), 117, 203
  filters, 207–210, 213, 257
  floating, 209
Frequency-emphasizing network (FEN) filters, 151–154, 156, 160, 252
Friend (SAB) circuit, 178

Gain-bandwidth product, 222, 230–235, 258, 266
Gain-sensitivity product, 6, 242–243
Gaussian characteristic (*see also* Approximation, delay), 20–21
General differential-input filters, 174–176, 181, 183
GIC (generalized impedance converter), 107–108, 115–117, 123
  conversion factor, 206
  current type (CGIC), 107, 116
  simulation of floating inductance, 190, 213
  simulation of grounded inductance, 117, 186, 213
  transmission matrix, 202–203
  voltage type (VGIC), 107
GIC-derived circuit, 210–211, 213, 216, 217–218, 244, 247, 250
GIC methods, 202–211
Graph determinant, 270
Graph, planar, 31
Group delay, 19, 232
Gyrator, 6, 109, 112–114, 213
  analogy with gyroscope, 109
  Antoniou, 114, 188, 216
  conductance of, 112–113, 216
  conductance mismatch, 190
  flotation circuit, 189
  fully-floating, 113–114, 189
  grounded, 112, 189
  matrix, description of, 110, 112, 190
  nonideal model, 186
  n-port, 117–118
  semi-floating (quasi-floating), 189
  simulation of inductance, 112, 185–190, 212
  three-terminal, 112–113, 186–189
  two-amplifier, 189

Higher-order filters, 6, 72–74, 193–199, 210, 212, 218, 251–254, 258
High-frequency realizations, 50, 109, 142, 185, 232, 235, 236, 243–244, 258
Highpass filters, 129, 132, 140, 143, 157, 168, 170, 185, 191
Highpass notch, 178
Highpass response, 21, 22, 23
High-Q realizations, 133, 153–154, 166–168, 117, 189, 231, 236, 254
Horowitz decomposition, 64–65
Hurwitz polynomial, 34, 64

Immittance:
  converters, 104–109
  functions, 26–29, 39, 67
  inverters, 104–109, 112–114
Impedance level, 202, 205, 233, 252
Imperfections (*see* Component, Operational amplifier)
Inductive-π arrangement, 190, 204–205
Inductor, practical, 54–55, 185
Inductor simulation:
  direct, 185–193, 212
  grounded, 185–188, 213
  floating, 189–191, 209, 213
  quality of, 186
  semi-direct, 192
  using GICs, 190
  using gyrators, 112, 186–190, 212
  using multi-port circulator, 190–191, 192
Inductor subnetwork, 203–207, 213
Integrated circuits (microelectronic), 47, 97, 221–225, 233, 245–246, 278
  electron-beam lithography, 258
  hybrid, 221, 227
  monolithic, 2, 221
  production processes, 227–228
  substrate-area requirements, 245–247
  thick-film, 161, 226–227, 258, Plate 2
  thin-film, 47–48, 117, 225–226, 228, Plate 1
Integrators, 193, 195, 199, 200
  dual-input, 197
Inverters:
  impedance, 104–109, 112–114, 123
  negative (NII), 108
  positive (PII) (*see* Gyrator)
  multi-port, 117–118

Jump resonance, 233, 236

Laboratory filters, 245
Ladder network:
  bridged, 45–46
  Cauer, 38
  cumulant (continuant), 44–45
  general n-stage, 45
  interactive effects in, 193
  lowpass, 38, 42, 194, 196, 208, 210
  parallel, 40–41
  RC, 39
  three-section, 45, 46
LC network: filters, 1, 26, 185, 189
  normalized lowpass prototypes, 191–192
  one-port properties, 26–28
$LC:RC$ transformation, 29–30
LCR passive resonator, 157, 197
Leapfrog filter, 6, 193–197, 212, 213, 218, 254
Legendre polynomials, 14
Lim-circuit (controlled positive feedback):
  bandpass, 160–161
  notch, 134–135, 256

INDEX

Low-frequency realizations, 244–245
Lowpass filter, 3, 157, 161, 170, 191–192, 194, 210, 235, 252, 256, 279, Plate 2
Lowpass notch, 178
Lowpass response, 10–16

Mason's General Rule, 139, 199, 270
Matrix conversions, 36, 272
Midband gain (*see also* Centreband gain), 154, 254
Minimum-phase transfer function, 10, 39
Mixed-source unity-gain circuit, 148–151, 156, 160, 244, 246, 250, 258
Monte Carlo analysis, 82
Multi-channel filters, 256
Multiparameter statistical sensitivity function, 80, 242–243, 258
Multiple-loop (negative) feedback filter (MLFB), 168–172, 179, 180, 181, 182, 231, 243, 246, 251, 256
 modified form, 177–179, 180, 181
Multiple-loop systems, 6, 253–254, 258
 follow-the-leader feedback (FLF), 253
 generalized FLF, 253
 inverse FLF, 254
 leapfrog, 193–197, 218, 254
 primary resonator block (PRB), 254
 shifted-companion form, 254

NIC (negative impedance converter) (*see also* Converter, impedance), 2, 105–106, 109–112, 123
 realization sensitivities, 67
n-port network, 203–207
Nodal admittance matrix, 267–269
Noise, 4, 250, 251, 253
Nonlinear effects, 235, 243, 258
Norator (*see* Pathological elements)
Normalization, 263
Notch filters, 50, 174, 178, 210, 235, Plate 1
 controlled-source floating twin-T, 133–134, 137–138
 controlled-source grounded twin-T, 135–138, 257
 null-adjusted, 41, 47, 50
Notch function (*see also* Bandstop response), 40–41, 47, 57, 168, 169
Nullator (*see* Pathological elements)

One-port RC and LC properties, 26–29
 realizability, 32–34
Operational amplifier:
 voltage, 2, 6, 85, 87–91, 172, 221, 222, 225
 common-mode rejection ratio (CMRR), 229
 compensation, 230, 239
 differential-input admittance, 87
 differential input type (DIOA), 87–91, 172, 177, 193, 269
 differential-output type (paraphase), 87, 89, 91
 dominant pole, 230
 field-effect input, 98, 221
 filter realizations using, 163–184, 186, 276
 general 5-terminal, 87–89
 ideal, 85, 87, 123, 164, 172, 174, 178
 imperfections (nonidealities), 228–229, 263
 large-signal bandwidth, 235, 258
 linear models, 87–89, 90–91, 229
 nonlinear model, 235
 open-loop frequency response, 229–230
 open-loop gain, 70, 89, 157, 222, 228–230
 output admittance, 87, 89, 95
 output impedance (single-ended), 222, 228–229
 pathological equivalent, 88–89
 single-input, 88–89, 163–172, 269
 slew rate, 235–236, 243, 252
 types (commercially available), 95, 222, 278
Optimum design (and comparisons), 241–247
 of controlled-source realizations, 157–161
 of single-OA realizations, 179–183
 of analogue-simulation realizations, 212–218
Oscillators, 4, 55, 257

Parallel-ladder synthesis, 40–41, 164
Parasitic elements, 54–55, 186–187, 217, 236–237
Passband gain, 131, 171
Passive networks (*see also* LC, LCR and RC networks):
 ladder, 44–45
 limitations of, 1, 50, 52
 LLFBN, 26, 32
 one-port, 26–29, 37–39
 pole positions, 28
 realizability, 32–37
 two-element-kind frequency characteristics, 27
 two-port, 35–37, 39–47
 unbalanced, OCVTRs for, 46
Pathological network elements, 85–87, 112, 123, 258
 norator and nullator, 85–86
 nullator-norator equivalents, 86–87, 89, 92, 93–94, 108, 111, 112, 113–114
 nullor, 85–86, 113–114, 123
Peak gain (*see also* Centreband gain), 201
Perspective charts:
 active elements, 123
 analogue-simulation realizations, 213
 controlled-source realizations, 156
 selectivity and sensitivity, 81
 single-OA realizations, 180
'Phantom' zeroes, 67
Phase-locked loop, 4, 251, 259
PIC (positive-impedance converter), 106–107

283

INDEX

Pole-zero cancellation, 47, 152, 168
Positive-real functions (PRFs), 34
Power dissipation, function, 33–35, 54
   of filter components, 222, 241, 246
Power supplies (d.c.), 87, 189, 239, 241
Predistortion, 173, 247, 265–266
Private pole, 35

Q factor (*see also* Selectivity, and High-Q realizations), 55–56, 158, 181, 214–215
   control (of GIC-derived filter), 211
   control (of modified MLFB filter), 177, 183
   control (of single-feedback filter), 168
   enhancement (of positive-gain single-source filter), 133
   perturbation, 231–232
   tuning, 247–251
Q sensitivity, 69–70, 138, 157, 159, 160, 179–183, 212, 215, 216–217
Quality factor:
   of coil inductance, 54–55
   of simulated inductance, 112, 114, 187, 189

RC-embedded filters (dual source), 143–147, 156, 158–159, 160, 232
RC networks, passive, 28–30
   bridged twin-T, 167
   continued-fraction synthesis, 38–39
   driving-point synthesis, 37–39
   ladder, 174
   parallel-ladder synthesis, 39
   transfer-function synthesis, 39–41
   twin-T, 45–47, 134
Realizability, 9, 10, 26, 32–37
   driving point, 33–34
   residue conditions, 35
   transfer function, 35–37
Reciprocity, 26, 109
'Rejectivity', 53, 78
   gain-slope, 56–57, 78
   inverted-half-power, 57
   phase-slope, 57
Reliability, 245
Resistance per square (ICs), 223, 226
Resonator, active, 138–139
   LCR, 157
Response (frequency), ideal lowpass, 9–10
   practical bounds, 10, 82, 273–274
Riordan gyrator, 117
RL networks, 29

Sallen and Key circuit (positive-gain single-source), 125, 126, 128–133, 157, 231, 246, 256
Schoeffler sensitivity criterion, 79, 82
Selectivity (*see also* Q factor), 5, 52–59, 81
   energy formulation, 33, 54
   factor, 58–59
   half-power ('classical' Q), 53

notch response (*see also* 'Rejectivity'), 56–57
second-order function (Morris Q), 56
Sensitivity, 2, 5, 33, 52, 59–82
   'classical' (relative), 60, 61
   coefficient, 62
   comparisons for different filters, 157–161, 179–183, 212–218, 242–243, 254
   differential, 60
   frequency, 69, 70, 157–161, 182, 218, 242–243
   gain, 61
   high-order, 72–74, 217, 218
   incremental relative, 75
   large deviation, 75–76
   minimization, 63–68, 70–72, 79, 252
   multiparameter, 76–82
   optimal decompositions for, 64–68, 109
   parameters, 216
   passive, 157, 199
   phase, 61
   pole-position (root), 62
   pole-zero, 62
   Q, 69, 70, 138, 157–161, 179, 182, 183, 216–217, 242–243
   relative-pole, 74
   second-order, 68–70
   single-parameter, 59–81
   statistical multiparameter, 5, 80, 242–243, 258
   transfer function, 69, 216
   worst-case, 79
Short-circuit current gain, 87
Signal-flow graphs, 270–271
   of multiple-loop systems, 253
   of second-order system, 198–199
   of state-variable systems, 198–199
Simulated-inductance circuits, 6, 191–197, 204, 206, 208, 210, 218, 245, 248, 252, 257
Simulated inductance: grounded, 185–189, 212
   floating, 189–191, 212
Simulation of inductor subnetworks, 203–207
Single-amplifier biquadratic (SAB) section, 177–178
Single-feedback circuits, 163–168, 180
Stability, 52, 59–60, 130–133, 156
   absolute, 60, 69, 133
   limits for controlled-PFB filters, 130–131, 133, 137–138, 141
   marginal, 59
   of operational amplifier, 229–230
   open-circuit, 60, 115
   short-circuit, 60, 115
   strict, 59, 229–230
STAR configuration, 178
State-variable filter, 6, 197–202, 213, 244, 247–248, 250, 252, 256
State voltage, 198

INDEX

Statistical sensitivity function, 80, 242–243
Synthesis, passive network, 5, 9, 26, 37, 152, 163, 263, 265
   driving-point function, 37–39
   transfer function, 39–41
Synthesis, active network (*see also* Co-efficient-matching technique):
   FDNR, 207–210
   FEN, 152
   general differential-input circuits, 174–175
   single-controlled-source, 125–127
   single-feedback circuits, 163–168
   single-operational-amplifier, 179–180
   with optimum-sensitivity decompositions, 64, 70

T networks, RC, 41, 42–43, 128, 132, 167
Temperature coefficients, 223, 224, 240
   effects, 6, 52, 239–241
   tracking, 246
Test filters, 256–257
Thick films, 161, 223, 226, 258, Plate 2
Thin films, 47, 117, 168, 223, 224, 225–226, 228, Plate 1
Tolerance limits, 80
Topology, 41, 270
Transadmittance (transconductance) amplifier, VCCS, 88, 93, 97–99
   multiterminal, 102–104
Transfer admittance, 41, 42–43, 44, 88, 126, 164
Transfer function (*see also* Realizability, Sensitivity, Synthesis), 9
   amplitude approximation, 10
   delay, 19
   second-order, 19, 21, 41, 47, 55, 57, 68, 126
   third-order, 47, 72–74, 168
   n'th-order, 198
Transfer impedance, 101, 153
Transimpedance amplifier, CCVS, 88, 93, 101
Transmission parameters (of differential OA), 87–88
   (of active elements), 123
Transformations:
   complex impedance (FDNR), 207
   frequency, 21–24, 274–275
   $LC:RC$, 29, 39
   lowpass-to-highpass, 132, 191
   $RC:CR$, 30
Transformer, ideal (IT), 106–107, 114–115

Transistors, bipolar, 2, 4, 92, 98, 100, 221, 236
   emitter-follower stages, 142, 228, 244
   field-effect, 92, 221, 233
   pathological model of, 86
Trimming, laser, 2, 4, 161, 227, 246, 248–250, Plate 2
Tuning, 6, 160, 168, 202, 211, 220, 246, 247–251, 256, Plate 2
   automatic, 250–251
   deterministic, 249–250
   digital, 256, 257
   functional, 247–248
   high-order, 254
   twin-T filters, 47, 134, 160, 248
Twin-T RC network, 40, 41, 45–47
   active configurations, 133, 137–138, 147, 161, 177, 183, 244, 246
   balanced, 154
   bridged, 166–168. 173, 183
   bridged-balanced, 43, 167, 173
   floating, 133–134
   null-adjusted, 134, 168
   rotated, 147
   selectivity, 47
   sensitivity, 183
   tuning, 47, 134, 161, 248
Two-integrator-loop (state-variable) filter, 212
Two-port network, 32, 105
   admittance parameters, 35, 36–41, 42–43, 49, 163, 272

Unbalanced network, 36, 39, 143, 162
Universal active filter (two-integrator-loop), 200–202, 212, 214–215, 217, 256

Variability, statistical, 80
Variation, relative, 75, 231, 242
   multiparameter, 76–77
Virtual earth (ground), 85, 164, 169
Voltage follower, 97, 143, 154
Voltage transfer ratio, 36, 45, 46, 126, 127, 128, 164, 169, 211
Voltage transfer function, open-circuit, 35–36
Voltage amplifier (VCVS), 92–97, 125–147
   unity gain (GVUGA), 96, 148

Wien-bridge circuit, 177

Zeroes:
   of transfer function, 39
   'phantom', 67